Polyamines in Fungi

Their Distribution, Metabolism, and Role in Cell Differentiation and Morphogenesis

MYCOLOGY SERIES

Editor
J. W. Bennett
Professor
Department of Plant Biology and Pathology
Rutgers University
New Brunswick, New Jersey

Founding Editor
Paul A. Lemke

1. *Viruses and Plasmids in Fungi*, edited by Paul A. Lemke
2. *The Fungal Community: Its Organization and Role in the Ecosystem*, edited by Donald T. Wicklow and George C. Carroll
3. *Fungi Pathogenic for Humans and Animals (in three parts),* edited by Dexter H. Howard
4. *Fungal Differentiation: A Contemporary Synthesis*, edited by John E. Smith
5. *Secondary Metabolism and Differentiation in Fungi*, edited by Joan W. Bennett and Alex Ciegler
6. *Fungal Protoplasts*, edited by John F. Peberdy and Lajos Ferenczy
7. *Viruses of Fungi and Simple Eukaryotes*, edited by Yigal Koltin and Michael J. Leibowitz
8. *Molecular Industrial Mycology: Systems and Applications for Filamentous Fungi*, edited by Sally A. Leong and Randy M. Berka
9. *The Fungal Community: Its Organization and Role in the Ecosystem, Second Edition*, edited by George C. Carroll and Donald T. Wicklow
10. *Stress Tolerance of Fungi*, edited by D. H. Jennings
11. *Metal Ions in Fungi*, edited by Güfcnther Winkelmann and Dennis R. Winge
12. *Anaerobic Fungi: Biology, Ecology, and Function*, edited by Douglas O. Mountfort and Colin G. Orpin
13. *Fungal Genetics: Principles and Practice*, edited by Cees J. Bos
14. *Fungal Pathogenesis: Principles and Clinical Applications*, edited by Richard A. Calderone and Ronald L. Cihlar
15. *Molecular Biology of Fungal Development*, edited by Heinz D. Osiewacz
16. *Pathogenic Fungi in Humans and Animals, Second Edition*, edited by Dexter H. Howard
17. *Fungi in Ecosystem Processes*, John Dighton
18. *Genomics of Plants and Fungi*, edited by Rolf A. Prade and Hans J. Bohnert
19. *Clavicipitalean Fungi: Evolutionary Biology, Chemistry, Biocontrol, and Cultural Impacts*, edited by James F. White Jr., Charles W. Bacon, Nigel L. Hywel-Jones, and Joseph W. Spatafora
20. *Handbook of Fungal Biotechnology, Second Edition*, edited by Dilip K. Arora
21. *Fungal Biotechnology in Agricultural, Food, and Environmental Applications*, edited by Dilip K. Arora
22. *Handbook of Industrial Mycology*, edited by Zhiqiang An
23. *The Fungal Community: Its Organization and Role in the Ecosystem, Third Edition*, edited by John Dighton, James F. White, and Peter Oudemans
24. *Fungi: Experimental Methods in Biology*, Ramesh Maheshwari
25. *Food Mycology: A Multifaceted Approach to Fungi and Food*, edited by Jan Dijksterhuis and Robert A. Samson
26. *The Aspergilli: Genomics, Medical Aspects, Biotechnology, and Research Methods*, edited by Gustavo H. Goldman and Stephen A. Osmani
27. *Defensive Mutualism in Microbial Symbiosis*, edited by James F. White, Jr. and Mónica S. Torres
28. *Fungi: Experimental Methods In Biology, Second Edition*, Ramesh Maheshwari
29. *Fungal Cell Wall: Structure, Synthesis, and Assembly, Second Edition*, José Ruiz-Herrera
30. *Polyamines in Fungi: Their Distribution, Metabolism, and Role in Cell Differentiation and Morphogenesis*, Laura Valdés-Santiago and José Ruiz-Herrera

Polyamines in Fungi

Their Distribution, Metabolism, and Role in Cell Differentiation and Morphogenesis

Laura Valdés-Santiago
Centro de Investigación y de Estudios Avanzados del Instituto Politécnico Nacional
Irapuato, Gto. México

José Ruiz-Herrera
Centro de Investigación y de Estudios Avanzados del Instituto Politécnico Nacional
Irapuato, Gto. México

CRC Press is an imprint of the
Taylor & Francis Group, an **informa** business

CRC Press
Taylor & Francis Group
6000 Broken Sound Parkway NW, Suite 300
Boca Raton, FL 33487-2742

First issued in paperback 2019

ISBN-13: 978-1-4987-1742-7 (hbk)
ISBN-13: 978-0-367-37710-6 (pbk)

Library of Congress Cataloging-in-Publication Data

Valdes-Santiago, Laura, author.
Polyamines in fungi : their distribution, metabolism, and role in cell differentiation and morphogenesis / Laura Valdes-Santiago and Jose Ruiz-Herrera.
pages cm. -- (Mycology ; 30)
Includes bibliographical references and index.
ISBN 978-1-4987-1742-7
1. Polyamines. 2. Fungi. I. Ruiz-Herrera, Jose, author. II. Title. III. Series: Mycology series ; v. 30.

QP801.P638V35 2015
579.5--dc23 2015022573

Visit the Taylor & Francis Web site at
http://www.taylorandfrancis.com

and the CRC Press Web site at
http://www.crcpress.com

To my beloved husband and intellectual partner, Dr. José Luis Castro-Guillén, his comments, support, and encouragement (that all writers need) were essential to the culmination of this book, and to our beloved kids, Josué and Jacob, who make it all worthwhile.

A special thanks to my mentor, Dr. José Ruiz-Herrera, for entrusting me with the uplifting task of being part of this book, as well as for being my professor during my academic life.

Laura Valdés-Santiago

With love to my wife, daughters, sons, granddaughters, and grandsons, for all the happiness they have brought to my life.

José Ruiz-Herrera

Contents

Preface

It is interesting to point out that polyamines were originally identified in a variety of different sources: human and animal sperm, different animal organs, as products of bacterial activity on decomposing corpses, etc., with apparently nothing in common, except their general chemical structure and as mere chemical curiosities. This obscure origin, quite different from the interest awakened by sugars, amino acids, vitamins, and nucleotides, in no way anticipated the essential role of polyamines in all living organisms. It was not until recent years that the study of polyamines, their mechanisms of synthesis, and the roles they play in metabolism has flourished, becoming a fertile field of intense research.

Encouraged by the information generated in our laboratory by many years of studying the roles of polyamines in fungal differentiation, as well as the many unanswered questions surrounding these molecules, we decided to write this book, moreover considering that no extensive review focused on polyamines in fungi has been recently written. Accordingly, the aim of this book is to provide an up-to-date account of the contribution and the potential use of fungi as models for the basic study of polyamines, in comparison with other organisms.

Of particular importance are the unanswered questions regarding polyamines that are explored, although unfortunately answered only in part. Among them, we may cite, are as follows: What is the individual function of each one of the polyamines; what is the answer to their high levels of regulation, including transport and back conversion; what is the precise role of spermine, taking into consideration that it is present only in some organisms, including several fungal groups, while most of them only possess putrescine and spermidine; and what are the basic mechanisms of function of polyamines in all the roles they have in the cells, including differentiation?

We have tried to write all the chapters in which the book is divided in the form of self-contained reviews of specific points, but in such a way that they invite the reader to go through the whole book. Nothing would be more satisfactory to us than fulfilling this expectancy.

Laura Valdés-Santiago
José Ruiz-Herrera
Irapuato, México

Acknowledgments

We extend our thanks to our students and collaborators who helped to obtain the experimental results described in the text; to Centro de Investigación y de Estudios Avanzados of Instituto Politécnico Nacional, for the facilities received; and to Consejo Nacional de Ciencia y Tecnología (CONACYT) for the partial support to our research.

We are also extremely grateful to Dr. José Luis Castro-Guillén for the preparation of the illustrations of the book.

Authors

Laura Valdés-Santiago has been a postdoctoral researcher at the Department of Biotechnology and Biochemistry of the Center for Research and Advanced Studies of the National Polytechnic Institute in Irapuato, México, since 2013.

She obtained a bachelor of science in biochemical engineering from Universidad Autónoma Metropolitana (Mexico City). Most of her academic career was spent at the Center for Research and Advanced Studies of the National Polytechnic Institute in Irapuato, México, where she obtained her master's of science in 2008 and her doctorate in plant biotechnology in 2011. In the same year, she joined the Department of Genetic Engineering as a visiting researcher. Her research interest has been focused on polyamine metabolism in fungi, a topic on which she has published several original and review papers.

José Ruiz-Herrera is professor emeritus at the Department of Genetic Engineering of the Center for Research and Advanced Studies of the National Polytechnic Institute in Irapuato, México.

Born in México City, he got his professional degree in microbiology at the National Polytechnic Institute in México City, and his doctorate in microbiology from Rutgers, The State University of New Jersey, United States.

Dr. Ruiz-Herrera has been professor and chairman of the Microbiology Department of his Alma Mater, founder of the Institute of Experimental Biology of the University of Guanajuato, and visiting professor at the University of California, San Diego and Riverside campus, and the Spanish universities of Valencia, Salamanca, Sevilla, and Extremadura.

He has been the recipient of important awards, including the award for young investigators from the Mexican Academy of Sciences, the Ruth Allen Award from the American Phytopathological Society, Miguel Hidalgo Award from the State of Guanajuato, and the National Award in Sciences from the Mexican government.

He has served on many scientific committees and as reviewer and associate editor for several prestigious journals. He is author of more than 250 original papers and reviews in international journals and four books and was editor of two books. He has directed over 25 master's theses and 36 doctorate dissertations. His current research interests are the regulation of development and differentiation in fungi, the synthesis and assembly of the fungal cell wall, and polyamine metabolism in fungi.

CHAPTER 1

Introduction

Polyamines are aliphatic amines that, at physiological values of pH, have a positive charge. Polyamines are widely distributed in nature and are essential for the growth of all living organisms because of their fundamental role as modulators of a large number of physiological processes. Interestingly, the importance that their original chemical characteristics, the different sources in which they were found or isolated, and even their common names have received in modern times was not anticipated. The original reference of polyamines is very old. Accordingly, spermine crystals were first observed by "The Father of Microbiology" Antonie van Leeuwenhoek in 1678 by the microscopic analysis of human seminal fluid, from where the name given to the molecule was derived more than two centuries afterward by Ladenburg and Abel (1888). Another polyamine, spermidine, was isolated and then chemically synthesized from bovine pancreas (Dudley, Rosenheim, and Starling 1927) by the same authors that firstly synthesized spermine (Dudley, Rosenheim, and Starling 1926). The third important polyamine, putrescine, and a more rare polyamine, cadaverine, were described as products of the bacterial action during flesh decomposition, hence their names. The first record of putrescine dates from 1885, when Ludwig Brieger described it as a molecule presented in putrefying material. It is obvious that these odoriferous molecules come from such different sources and have in common only their chemical similarities: Polyamine aliphatic molecules (see their structures in Figure 2.1 of Chapter 2) conceivably were considered rarities with no common or important physiological activities.

However, the essential role of polyamines in growth was deduced a long time after all these seminal studies that led to their isolation from natural sources, the deduction of their chemical structure, and their chemical synthesis. It was not until 1970 when their essential role in the fungus *Neurospora crassa* was deduced by the demonstration that the phenotype of an auxotrophic mutant of the fungus was reverted by addition of spermidine, putrescine, or its precursor (see discussion further in the chapter), ornithine (Davis, Lawless, and Port 1970), to its growth medium. Table 1.1 shows a recompilation of the dates of the important discoveries regarding polyamines, with emphasis on fungi.

As already indicated, polyamines are aliphatic amines that are synthesized by both prokaryotic and eukaryotic cells. The most common ones are putrescine

Table 1.1 Historical Account of Polyamine Research with Emphasis in Fungi

Author	Contribution	Reference
Antonie van Leeuwenhoek	Observation of spermine crystals in 1678	Ladenburg and Abel (1888)
Ludwig Brieger	Isolation of putrescine	Ludwing (1885)
Ladenburg and Abel	Isolation of spermine from human sperm	Ladenburg and Abel (1888)
Dudley, Rosenheim, and Starling	Synthesis of spermine	Dudley, Rosenheim, and Starling (1926)
Dudley, Rosenheim, and Starling	Isolation and synthesis of spermidine from bovine pancreas	Dudley, Rosenheim, and Starling (1927)
Sakurada	Determination of an effect of spermine on fermentation by *Saccharomyces cerevisiae*	Sakurada (1962)
Castelli and Rossoni	Relationship between polyamines and nucleic acids in yeast	Castelli and Rossoni (1968)
Davis, Lawless, and Port	Discovery of the essentiality of polyamines in *Neurospora crassa*	Davis, Lawless, and Port (1970)
Garcia; Inderlied, Cihlar, and Sypherd	Differentiation of *Mucor racemosus* related with alteration in polyamine levels	Garcia et al. (1980); Inderlied, Cihlar, and Sypherd (1980)
Fonzi and Sypherd	First isolation of a gene involved in polyamine synthesis	Fonzi and Sypherd (1985)

(butane-1,4-diamine), spermidine [*N*-(3-aminopropyl)butane-1,4,diamine], and spermine [*N*-*N*′-bis(3-aminopropyl)butane-1,4,diamine].

The physiological significance of polyamines can be deduced from the ubiquity of these compounds, both in microorganisms and throughout the plant and animal kingdoms. Although spermidine and spermine have been the polyamines considered most important in higher eukaryotes (Tabor and Tabor 1984), this does not occur in lower eukaryotes and bacteria, where putrescine is considered very important and spermine is lacking in bacteria and most fungi.

The cationic nature of polyamines enables them to interact with negatively charged molecules, such as nucleic acids, negatively charged lipids, and some proteins, but besides this rather unspecific binding, a plethora of experiments have demonstrated the specificity of polyamine interactions with different biological compounds of the cell. As an example, proton nuclear magnetic resonance (H-NMR) studies demonstrated that spermine and other polyamines bind to transfer ribonucleic acid (tRNA), but with a ratio of one or two molecules of polyamine per tRNA molecule, despite the fact that the RNA molecule contains a large number of negative charges, demonstrating a specific type of recognition by these molecules. Moreover, it has been demonstrated that through this binding, polyamines control the correctness and effectiveness of the translation process (Frydman et al. 1992; Igarashi and Kashiwagi 2000, 2006). *In vitro*, it has been demonstrated that physiological concentrations of polyamines increase DNA replication, transcription, and mRNA translation (Frugier et al. 1994). *In vivo*, the intracellular concentration of polyamines is related with cell proliferation (Heby 1981).

Polyamine concentration in the cell is subjected to a complex set of regulatory mechanisms, and recent discoveries have shown that each one of the polyamines may accomplish specific functions.

In the case of fungi, the first reference to some role of polyamines in their metabolism was probably the communication from Sakurada (1962), who reported an effect of spermine over sugar fermentation by baker's yeast. Afterward, yeast served as a model system to establish a relationship between polyamines and nucleic acids (Castelli and Rossoni 1968; Heby and Aqurell 1971). Meanwhile, in *N. crassa*, after the demonstration of the essentiality of polyamines (see previous discussion), the levels of spermidine and spermine were correlated with the concentrations of magnesium and the rate of RNA and DNA synthesis (Viotti et al. 1971), and the authors suggested that polyamines could partially substitute for the role of magnesium.

Interestingly, fungi have demonstrated to be excellent model systems to elucidate some specific roles of polyamines. These organisms are ideal models of analysis, since they are among the simplest eukaryotic organisms; are easy to grow under controlled conditions in the laboratory; are, in large quantities, able to be subjected to biochemical analyses; and are amenable to the operation of all the modern genetic and molecular tools. Some of the documented functions of polyamines in fungi can be extended to higher eukaryotes. Among them, we may cite the seminal studies by the group of Sypherd on the role of polyamines on the important phenomenon of cell differentiation in fungi, through their demonstration of the alterations of polyamine levels during the dimorphic transition (dimorphism is the capacity of a fungus to grow in the form of yeast or mycelium depending on the environmental conditions) of *Mucor racemosus* (Garcia et al. 1980; Inderlied, Cihlar, and Sypherd 1980). These results were followed by the demonstration that monomorphic yeast mutants (i.e., that they cannot grow in the form of mycelium and grow only in the yeast form) of *Mucor bacilliformis* contained reduced levels of ornithine decarboxylase, the first enzyme of the biosynthetic pathway of polyamines (Calvo-Mendez, Martinez-Pacheco, and Ruiz-Herrera 1987), and by the demonstration that the involvement of polyamines in the differentiation phenomenon of fungal dimorphic transition could be extended to other unrelated species, such as *Ustilago maydis* (Guevara-Olvera, Xoconostle-Cazares, and Ruiz-Herrera 1997), *Yarrowia lipolytica* (Guevara-Olvera, Calvo-Mendez, and Ruiz-Herrera 1993; Jimenez-Bremont, Ruiz-Herrera, and Dominguez 2001), and others. Other examples of the roles of polyamines on fungal differentiation are the following: spore germination (Trione, Stockwell, and Austin 1988; Rajam, Weinstein, and Galston 1989; Kummasook et al. 2013) and sporulation (Guzmán de Peña, Aguirre, and Ruiz-Herrera 1998; Bailey, Mueller, and Bowyer 2000). In the same way, fungi have been useful to analyze the role of polyamines in stress response (Valdes-Santiago, Cervantes-Chavez, and Ruiz-Herrera 2009; Valdes-Santiago and Ruiz-Herrera 2014) and mycelial growth (Herrero et al. 1999; Khatri and Rajam 2007).

Considering the importance of polyamines in all living organisms, and the particular advantages of fungi as experimental models, the aim of the present volume is to give an outlook of polyamine metabolism in fungi, including their distribution, physiological functions, and mode of action at the molecular level, as well as their

Table 1.2 Some Physiological Roles Reported for Polyamines in Eukaryotic Organisms

Roles in Fungi	Roles in Plants	Roles in Animals
Growth	Stress response	Cell proliferation
Dimorphism	Control of ion channels	Stress response
Spore germination	Embryogenesis	Membrane stabilization
Sporulation	Floral development	Prevention of diseases
Stress response	Senescence	Life span extension
Appressorium development		
Conidiation		
Life span extension		

mechanism of synthesis, in comparison with higher eukaryotes. In Table 1.2, we present some functions of polyamines discussed in this book. Phylogenetic analyses of the enzymes involved in their synthesis and their use as possible targets for the control of fungal diseases in plants and animals, as well as for other maladies, are also discussed.

CHAPTER 2

Chemistry and Cellular Localization of Polyamines

2.1 INTRODUCTION

Spermine phosphate was the first known naturally occurring polyamine observed in 1678 by Antonie van Leeuwenhoek. More than two centuries had to elapse before the structure of polyamines was determined (reviewed by Bachrach 2010). Afterward, the simpler diamine, putrescine, was discovered in dead animals in decomposition, from here came the origin of its name and the general feeling that this and its sister compounds were only molecules without any biological importance and mere products of the decay of the organic matter. Today, the structures of the most common polyamines present in living organisms: putrescine, spermidine, and spermine, are known as presented in Figure 2.1. They are molecules with two or more amino groups presenting positive charge at physiological pH. Natural polyamines have been well characterized, although as indicated in Chapter 1, they were considered of little importance for a long time, and their interest lagged behind that of other biologically important molecules, such as amino acids, sugars, and nucleotides. It was not until more recently that an ample number of physiological processes in which they are involved have been discovered.

Polyamines, in general, are derived from arginine, and their synthesis involves mainly the decarboxylation of ornithine by Odc (ornithine decarboxylase) to produce putrescine, although in plants and bacteria, there exists an alternative pathway to generate putrescine involving the enzyme Adc (arginine decarboxylase). Afterward, with the involvement of two aminopropyl transferases, putrescine is sequentially converted into the higher polyamines spermidine and spermine. These aminopropyl transferases are spermidine synthase (Spds) and spermine synthase (Spms). The first one utilizes putrescine as an acceptor for the synthesis of spermidine, while the second one uses spermidine as substrate to generate spermine. The aminopropyl groups are donated in both cases by decarboxylated *S*-adenosylmethionine, with *S*-adenosylmethionine decarboxylase (Samdc) being the enzyme responsible for SAM decarboxylation. On the other hand, there exists a back conversion mechanism of polyamines that starts with the acetylation of spermine, a reaction catalyzed by the enzyme spermine/spermidine-N^1-acetyltransferase (Ssat). Afterward, acetyl

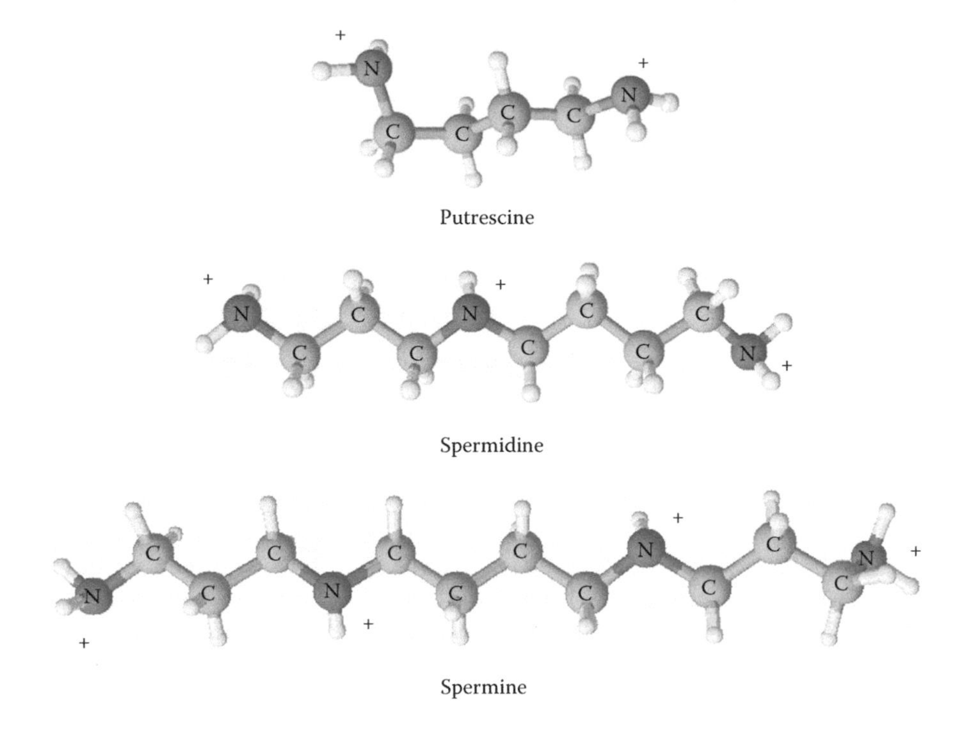

Figure 2.1 Three-dimensional representation of the most common natural polyamines. Carbon atoms are represented as spheres with the letter C, nitrogen atoms as spheres with the letter N, and hydrogen atoms as white small spheres. Protonation sites are indicated with the + symbol.

spermine becomes the substrate for polyamine oxidase (Pao) to produce spermidine. The action of the same acetyl transferase and Pao is responsible for the transformation of spermidine into putrescine.

An important factor for the biological functions of polyamines is their high degree of positive charges and their interaction with negative molecules, as will be discussed with more detail later in this chapter. Modulation of specific DNA–protein interactions by polyamines has been suggested to be a means to regulate gene expression, as confirmed by sound evidences showing that polyamines may regulate enzyme synthesis through their interactions with nucleic acids (Panagiotidis et al. 1995; Deng et al. 2000; Ouameur and Tajmir-Riahi 2004; Schuster and Bernhardt 2011); also polyamines are involved in the transcription of specific genes (Childs, Mehta, and Gerner 2003; Xiao and Wang 2011). Nevertheless, it has to be emphasized that to obtain further evidence of the mode of action of polyamines, knowledge of their exact localization in the cell is essential. In this chapter, we present highlights of some aspects of polyamines chemical properties that are important to gain information on how they behave in solution and their interactions with negatively charged biological molecules, as well as their repercussion on the biological activity of these molecules. And in addition, their subcellular localization will be discussed to understand how they affect cellular functions in general. Finally, we will analyze how the use of fungi as model systems may help to understand their mechanisms of action.

2.2 CHEMICAL STRUCTURE AND PHYSICOCHEMICAL CHARACTERISTICS OF POLYAMINES AND THEIR IMPACT ON THE INTERACTION WITH BIOMOLECULES

Polyamines are organic, polycationic, low-molecular-weight compounds that are widely distributed in nature. Structurally, polyamines are constituted by aliphatic chains with primary amino groups ($-NH_2$) on each extreme of the chain made of methylene groups ($-CH_2-$), and secondary amino groups (–NH–), located internally in the methylene chains. Branched polyamines present a similar arrangement (Morgan 1998). At least one of the three most common polyamines, putrescine (1,4-diaminobutane), spermidine (1,8-diamino-4-azaoctane), and spermine (1,12-diamino-4,9-diazadodecane) (Figure 2.1), are present in almost every living cell.

In aqueous solution, polyamines present a basic behavior, and hydrophobic interactions appear to be the source of binding energy with some molecules, and an example of this is the binding of polyamines that cause blocking of *N*-methyl-D-aspartate receptor ion channels (Cu, Bähring, and Mayer 1998). The polycationic nature of polyamines is the main factor that contributes to their physiological activity, since it has been reported that under physiological conditions, polyamines can interact with negative macromolecules such as DNA and RNA, inducing conformational transitions and affecting nucleic acids functions (Flink and Pettijohn 1975; Gosule and Schellman 1976; Ruiz-Chica et al. 2003; N'Soukpoe-Kossi et al. 2008).

Putrescine, spermidine, and spermine exhibit net charges close to +2, +3, and +4, respectively, as is shown in Figure 2.1, due to their primary amino groups found charged at physiological values of pH. The pKa values (the pH at which a functional group is 50% ionized) of their basic sites are proportional to the internitrogen distances. Consequently, the binding of polyamines to macromolecules could be explained by the pairwise formation of electrostatic interactions that depend on the spacing between the nitrogen atoms (positive-ionizable groups) in the polyamine chain. This, in turn, influences both the base strength and conformational flexibility of the molecule (Ganem 1982). In the case of polyamine–nucleic acid interactions, these factors are responsible for the electrostatic binding with the phosphate groups of DNA and its induction of the conformational change (Basu et al. 1990; Pastre, Pietrement, and Landousy 2006). After discussing the protonation constant of some amines and how the reduction in their basicity could be produced by the chain length separating their amino groups, Bencini et al. (1999) concluded that the fully protonated stage of the biogenic polyamine spermidine practically at neutral pH facilitates its well-known physiological role as charge neutralizer of DNA. Polyamines are the perfect compounds for creating zwitterionic structures, since the distance between amino groups of the three and four carbon atoms is practically the same as the distance between phosphate anions in the DNA backbone (Zarytova and Levina 2011).

The behavior of polyamines in solution is worth analyzing, considering that at the biological level, this is the way polyamines are present in the cell, and it is an important factor on how they interact with the anionic cellular components. It is well known that both the number of positive charges of polyamines and the distribution of charge along the molecule have important effects on its capacity to produce the DNA conformational changes mentioned previously and surely how they control their biological functions (Basu et al. 1990). Since many years ago, the need to understand the mechanisms by which putrescine, spermidine, and spermine mediate critical physiological processes has been established, and this necessity has led to substantial interest in establishing the physicochemical properties of these polyamines (Aikens et al. 1983).

A deepened study about the extent of ionization of the polyamines was developed by Blagbrough, Metwally, and Geall (2011). These authors reported the pKa values for each amine group in the important polyamines and pointed out the fact that for compounds with more than one ionizable center (atom or functional group), there is a pKa value for each center of ionization and stressed that the measurement of pKa data was crucial to understand their role in many biological processes (Blagbrough and Geall 2003; Blagbrough, Metwally, and Geall 2011). For example, when the effects of spermine and two analogs, 1, 11-bis (ethyl-amino)-4,8-diazaundecane and 1,19-bis(ethylamino)-5,10,15-trirazanonadecane, on the formation of nucleosomes *in vitro* were investigated, an alteration of nucleosome structure on newly replicated DNA was demonstrated (Basu et al. 1997). This effect was probably caused by the different distribution of charge of these analogs as compared with spermine.

2.3 BINDING OF POLYAMINES WITH CELLULAR MACROMOLECULES

As repeatedly observed, because of their chemical structure, polyamines interact with negatively charged macromolecules, forming an electrostatic binding. Numerous studies indicated that polyamines interact with nucleic acid, not only DNA but also RNA, and also with proteins and phospholipids (Bachrach 2005). These interactions are capable, as mentioned previously, to induce conformational transitions and affect the functions of these macromolecules. The question is whether the interactions are specific, based upon a specific site and geometry of the interactions, or nonspecific, based on charge alone. Several studies give evidences of nonspecific interactions with DNA (Burton, Forsen, and Reimarsson 1981; Braunlin, Strick, and Record 1982; Plum and Bloomfield 1990), but nevertheless, other authors concluded that polyamines may bind in a site-specific way, e.g., that polyamines bind preferentially to bent A-tracts and TATA sequence element (Matthews 1993; Lindemose, Nielsen, and Mollegaard 2005). It has been suggested that the DNA sequence for binding of polyamines should be specific to induce the union between DNA and transcription factors. However, Venkiteswaran, Thomas, and Thomas (2006) highlighted that the site- or base-specific interactions between polyamines and DNA are controlled by different factors such as hydrophobicity, electrostatic forces, and the presence of phosphate groups. The observation that spermidine inhibited *in vitro* DNA methylation by cytosine–DNA methyltransferases, but it did not affect the activity of the restriction enzymes that bind at the same sites, suggested the selective recognition of nucleotide sequences of DNA by polyamines (Ruiz-Herrera, Ruiz-Medrano, and Domínguez 1995). There are also *in vivo* evidences indicating that the effect of polyamines takes place at the level of DNA methylation (Karouzakis et al. 2012; Soda et al. 2013).

Early studies suggested that polyamines and histone acetylation function in concert to lower the stability and change the conformation of the nucleosome core, thus facilitating replication and transcription *in vivo* (Morgan, Blankenship, and Matthews 1987; Hobbs, Paul, and Gilmour 2002). In the same line, conformational changes of chromatin by polyamines have been reported (Matthews 1993). The strong affinity of polyamines for DNA has important biological implications (Feuerstein, Pattabiraman, and Marton 1990). Thus, the deletion of Gnc5p (a coactivator/histone acetyltransferase that induces relaxation of the nucleosome structure) from *Saccharomyces cerevisiae* affected the expression of genes such as *HO* and *SUC2*. Interestingly, polyamine depletion caused by the mutation of the gene encoding the ornithine decarboxylase (*ODC*) restored partially the expression of genes dependent on Gnc5p (Pollard et al. 1999). However, Hobbs, Paul, and Gilmour (2003) noticed that the prolonged deprivation of cellular polyamines led to the accumulation of cells in the G1 phase of the cell cycle, a reaction that could be responsible for the partial restoration of defects in the Gcn5p-dependent transcription. All these contradictory evidences indicate that, in general, how polyamines induce chromatin alteration is not clear. One possible explanation is that polyamines give rise to a more compact structure of chromosomal DNA that leads to entrapment of nucleosomes. Another possibility is that polyamines regulate genes encoding chromatin-remodeling proteins. Moreover, polyamines might interact with histones, affecting in this way the interaction of nucleosomes with DNA (Libby and Bertram 1980; Hobbs and Gilmour 2006).

Actually, it has been shown that biogenic polyamines cause DNA condensation in both isolated DNA (Patel and Anchordoquy 2006) and in chromatin (Todd et al. 2008). Putrescine, spermidine, and spermine are known to bind through the minor and major grooves of double-stranded DNA, while electrostatic attractions between these polycations and the DNA phosphate backbone also occur (Ouameur and Tajmir-Riahi 2004).

Excellent systems that can be used to evaluate the effect and the possible role of polyamines on chromatin structure are fungal models; for example, the determination of the precise position of nucleosomes *in vivo* is now possible in yeast (Brogaard et al. 2012; Wal and Pugh 2012). Genes involved in the biosynthesis of secondary metabolites are usually physically linked in the chromosome and constitute transcriptionally coregulated clusters in fungi. Accordingly, the close proximity of the genes would allow a coordinated transcriptional control by chromatin-based mechanisms (Gacek and Strauss 2012). In fact, *Aspergillus nidulans* is a potential model for *in vivo* structure/function analysis of histones, since its core histone genes have been cloned and sequenced and all but one are present in a single copy (Ehinger, Denison, and May 1990; Gonzalez and Scazzocchio 1997; Hays, Swanson, and Selker 2002).

Since, as repeatedly mentioned, the amino nitrogen of polyamines is completely protonated at physiological pH, polyamines are able to interact with phosphate groups because of the attraction of the opposite charges. *In vitro*, there is a cyclic structure formed by the intercalation of a phosphate anion between the *N*-terminal ends of two polyamines. These structures have been denominated nuclear aggregates of polyamines (NAPs), and it has been suggested that NAPs regulate the transition of DNA forms, favoring the Z-form, and also DNA protection (D'Agostino and Di Luccia 2002; D'Agostino, di Pietro, and Di Luccia 2006). Experiments with a monoclonal anti-Z-DNA antibody and spectroscopic techniques showed the ability of polyamines to induce the B-DNA to Z-DNA conformational transition. In this context, spermidine was found to be the most efficient inducer of the transition (Thomas and Messner 1988). Furthermore, the spermine molecule has been found to be located at the minor groove of the duplex, in a central GCGC DNA region, as shown by x-ray crystallographic analysis (Clark et al. 1990). However, Deng et al. (2000) mentioned that this result could be given because the oligodeoxyribonucleotides employed in these analyses were usually rich in GC. As indicated previously, other reports suggest that the interaction of polyamines with DNA appears to be specific and that putrescine and spermidine displayed preference for AT-rich regions (Schmid and Behr 1991). Recent studies have confirmed this last observation. Using some thermodynamics analyses such as circular dichroism spectroscopy and melting studies, it was shown that genomic AT-rich DNA displayed greater interaction with polyamines compared with genomic GC-rich DNA. Apparently, the length of the polyamine influences the force at which each one of them binds to DNA (Kabir and Kumar 2013).

In the same line, a base composition dependence of interactions of spermidine and spermine with genomic DNA from organisms with different GC content was determined by Raman spectroscopy. The results revealed that both spermidine and spermine resided along the major groove and not only interacted with phosphates but also were bound to the DNA backbone and perturbed the guanine N7 and thymine C5H3 sites (Deng et al. 2000). More important, Deng et al. (2000) observed the capacity of both spermidine and spermine to bind and condense genomic DNA while conserving the native B-form secondary structure. This observation is in opposition to the behavior of divalent metal ions that bind primarily to DNA bases and disrupt the B-DNA structure (Duguid et al. 1993; Duguid and Bloomfield 1995). Taking into consideration the importance of both phosphate and major groove sites in gene regulation (Steitz 1990; Pabo and Sauer 1992), the role of polyamines in gene regulation could be explained in the base of the binding or release of gene regulatory proteins given by the occupancy or modulation of major groove dimensions. In this context, both secondary structure and base composition would be important for polyamine recognition (Deng et al. 2000).

Most of the putrescine and spermidine content in *Escherichia coli* is bound to RNA, and a similar profile was reported in animal cells. This fact led some authors to suggest that polyamines may exert their effect through a change in the structure of RNA (Igarashi and Kashiwagi 2000). The physiological relevance of the RNA–polyamine interaction would be at the level of regulation of protein synthesis. It has been shown that polyamines have the ability to bind and influence the secondary structure of tRNA, mRNA, and rRNA. As a consequence, polyamines might modulate protein synthesis at different levels (Shimogori, Kashiwagi, and Igarashi 1996; Igarashi and Kashiwagi 2006; Ouameur, Bourassa, and Tajmir-Riahi 2010). It was observed that at low concentrations, the effectiveness of spermidine and spermine to bind to tRNA was higher than their ability to bind to DNA, therefore suggesting that at low concentrations, they might stimulate protein synthesis, rather than transcription (Wang et al. 1999; Ouameur, Bourassa, and Tajmir-Riahi 2010).

Due their cationic nature, it would be expected that polyamines interacted with negatively charged membrane components, affecting the properties of biological membranes. In this sense, it has been shown that protoplasts, spheroplasts, lysosomes, mitochondria, and erythrocyte membranes are stabilized in the presence of polyamines against stressful conditions, such as osmotic stress or lysis (reviewed by Schuber 1989). Furthermore, polyamines have the ability to form polyamine–phospholipid complexes. The consequences of this interaction are more than the simple neutralization of acid phospholipids. Taking into consideration the number of charges in polyamines, the distance between the charges, and their flexibility, these characteristics may confer to polyamines additional functions in their binding to phospholipids than those of inorganic cations (Tadolini 1988). The response of preformed membranes under electric voltage to treatment with different concentrations of polyamines resulted in their stabilization only with the addition of spermine, while putrescine and spermidine caused destabilization of the model phospholipid bilayers (Zheliaskova, Naydenova, and Petrov 2000). The biological implications of polyamines in their interaction with membranes are further discussed in Chapter 6.

In vitro, polyamines increased or inhibited the binding between several proteins and DNA, with spermine being more effective than spermidine and putrescine, probably because of their stronger cationic charge (Panagiotidis et al. 1995). Thus, the presence of polyamines increased the binding of the progesterone receptor isolated from rabbit uterus to DNA, and this binding was not affected by cytosolic components (Thomas and Kiang 1988). In breast cancer cells, it has been shown that polyamines increase the affinity of the nuclear transcription factor kappa B nuclear factor (NF-κB) with its DNA response elements (Zaletok et al. 2004). Some pathological conditions such as Parkinson's disease or Alzheimer's disease are induced by protein aggregation, and several evidences showed that polyamines are implicated in this process (Antony et al. 2003; Luo et al. 2013) or even may modulate protein aggregation (reviewed by Chowhan and Singh 2013). Moreover, while salts (KCl, NaCl) or amino acids such as arginine prevent, at certain concentrations, the aggregation of lysozyme *in vitro*, spermidine and spermine were found to be more effective in this capacity (Shiraki et al. 2003; Das et al. 2007; Ghochani and Moosavi-Nejad 2013). In this regard, it may be indicated that a stable neuronal cell line presenting a pathological polyglutamine aggregation phenotype related with the expression of a pathological-length polyglutamine protein of abnormal length (Q57) showed to be affected in the polyamine homeostasis compared with the cell line that expressed non-pathological-length polyglutamine protein (Q19) (Colton et al. 2004). All these data, some of them contradictory, suggest that at this point, there is the necessity for more *in vivo* assays to clarify polyamine specificity binding to nucleic acids and proteins to define the real physiological roles of polyamines.

Polyamine–protein interaction also has been studied in the context of polyamine uptake systems. A group of proteins identified in *Escherichia coli* and *Treponema pallidum* as transporters of extracellular polyamines contain multiple amino acid residues that function as polyamine binding sites. Regions that may be involved in the binding at the N^1 or N^6 of spermidine are conserved, as well as hydrophobic residues that could form Van der Waals interactions with polyamines (Sugiyama et al. 1996; Igarashi and Kashiwagi 1999; Machius et al. 2007). The existence of polyamine binding proteins in the plasma membrane of murine L1210 lymphocytic leukemia and human leukemia cells was discovered by the use of a photoaffinity labeling spermidine derivative containing a 4-azidosalicylic group at the N^4 position and a norspermidine photoaffinity labeling conjugated with 4-azidosalicylic acid at the N^1 position (Felschow et al. 1995). So far, in mammals, polyspecific cationic membrane transporters have shown to be involved in polyamine transport. Transporters such as OCT1 (members of the organic cation transporter family) recognize spermine, and evidences showed that spermidine may be also their substrate (reviewed by Abdulhussein and Wallace 2014). In the same sense, it is noteworthy that in membrane vesicles of rabbit small intestine and in human breast cancer cells, physicochemical parameters such as osmolality (the presence of inorganic salts such as NaCl or KCl), temperature, and extracellular pH affect putrescine and spermidine transport (Brachet, Debbabi, and Tome 1995; Poulin, Lessard, and Zhao 1995).

In the same way, the properties of polyamines as part of molecular complexes and coordination compounds with metals and nucleic acids based on noncovalent interactions of the amino groups should be taken into consideration, since the nature of their interactions, which are responsible for the biological activity of naturally occurring polyamines, changes under those conditions (Lomozik et al. 2005). Evidently, the association of polyamines with metal ions and inorganic cations is expected in the context of living cells, but whether these interactions would interfere or promote their biological activity remains unclear, hence the importance of the mentioned reports.

2.4 PHYSIOLOGICAL RESULTS OF THE INTERACTIONS OF POLYAMINES WITH CELL MACROMOLECULES AND STRUCTURES

As already discussed, there is no doubt of the interaction of polyamine with cell components possessing a negative charge. In fact, most of the polyamines of the cell are found bound to them. However, it is very important to define how these interactions affect growth, differentiation, and proliferation. Estimation of the concentration of polyamines bound to macromolecules has been reported in animal cells emulating physiological ionic conditions, showing that most of the spermidine and spermine were found bound to RNA and DNA (Watanabe et al. 1991). However, the physiological effects of polyamines and their metabolism depend on the physiological state of the organism. Accordingly, it would be a mistake to conclude that the documented specific complexes of polyamines, polyamine–RNA, polyamine–DNA, polyamine–protein, etc., exist in general in all systems and conditions. As already discussed previously, there exist examples showing the results that the association of polyamines with different macromolecules brings about. For example, accumulation of spermidine in *E. coli* inhibited protein synthesis, but it did not affected DNA and RNA synthesis at the stationary phase of growth, resulting in the loss of cell viability (Yoshida et al. 1999). Moreover, in this context, specific polyamine–RNA interactions have been described; thus, polyamines induce the adequate union between the Shine-Dalgarno sequence and the initiation codon in OppA mRNA (a gene encoding an oligopeptide permease) in *E. coli* (Yoshida et al. 1999; Klepsch et al. 2011). In the same way, spermidine modified the initiation region of OppA mRNA, interacting with the bulged out region, affecting its synthesis (Fukuchi et al. 1995; Higashi et al. 2008). The previously mentioned evidences supported the hypothesis of the mode of action of polyamine binding related with the increasing and/or the stimulation of transcription by frameshift or improved efficiency of the initiation codon (Kusama-Eguchi et al. 1991).

In contrast, treatment of FAO cells (from a hepatocarcinoma cell line) with difluoromethylornithine (DFMO), an inhibitor of Odc, induced a reduction in spermidine levels that was correlated with the inhibition of binding to DNA of the transcription factor AP-1. This inhibitory effect was reverted by spermidine addition, suggesting a specific effect of the polyamine. Interestingly, under the same conditions, another transcription factor STAT was not affected in its DNA binding by the presence of DFMO, suggesting that polyamines can be selective in affecting the binding of transcription factors to DNA (Desiderio et al. 1999).

However, opposite evidences have been also reported because polyamines can act indirectly as negative regulators at the posttranscriptional level. In mammalian cells, polyamine depletion induced the interaction of transcription factor-2 (ATF-2) with an RNA-binding protein (HuR), stabilizing the ATF-2 mRNA association and increasing in this way the levels of ATF-2 mRNA, and in opposition, the presence of polyamine presence repressed ATF-2 expression (Xiao et al. 2007). Interestingly, the effect of polyamine depletion in ATF-2 mRNA was not given by a possible action over the promoter, since cell treatment with DFMO did not affect its transcription, rather it was demonstrated that silencing of HuR reverted the stability of ATF-2 mRNA, indicating that polyamines control in some way the HuR levels in the cytoplasm (Xiao et al. 2007).

In eukaryotes, it is well known that genomic DNA is compacted in the nucleus in the form of chromatin. The structure of chromatin can control gene expression once it is altered in specific areas to allow the transcription of genes (Wu 1997; Li and Reinberg 2011). Polyamine–DNA interactions can take place at the chromatin level, as was discussed previously. A possible consequence of this interaction is chromatin compaction, as has been reported in HeLa cells, where high levels of spermidine or spermine caused charge neutralization of the sugar phosphate of the DNA backbone (Raspaud et al. 1999; Visvanathan et al. 2013). It has been reported that a biological implication of the interaction of polyamines with DNA in the chromatin is the modulation of gene expression (Balasundaram and Tyagi 1991; Childs, Mehta, and Gerner

2003; Zhu et al. 2012). At the level of interaction of polyamines with proteins, some evidences exist showing that polyamines have the ability to promote the binding of transcription factors with DNA response elements (Panagiotidis et al. 1995; Zaletok et al. 2004).

2.5 LOCALIZATION OF POLYAMINES IN THE CELLS OF DIFFERENT ORGANISMS, INCLUDING FUNGI

Studies on the cellular localization of polyamines and the enzymes involved in their synthesis have been made mostly in animal cells, followed by plants. Unfortunately, data on this issue are scant in fungi. Here, we review the data of localization in animal and plant cells with some detail and discuss the few results obtained with fungi, expecting that they may induce mycological researchers to become active in this area.

The spatial distribution and cellular localization of polyamines and the polyamine-synthesizing enzymatic machinery may help to better understand the functional roles of polyamines. Accordingly, it is necessary to identify the molecules or organelles to which polyamines are specifically linked, besides the knowledge we have that most of the cellular polyamines are bound to RNA (see previous discussion). For example, it has been indicated that in plants, amino oxidases (Mao's) are mostly enzymes colocalized with peroxidases, which may utilize the H_2O_2 derived from amine oxidation (Angelini, Federico, and Bonfante 1995). In agreement with these data, localization of Pao activity in rat tissues was found exclusively associated to the matrix of peroxisomes, using spermidine as substrate (Van den Munckhof et al. 1995). Interestingly, spermine oxidase (Smo) activity displayed nuclear and cytoplasmic localization (as is depicted in Figure 2.2) in animal cells. Probably participating in some other functions, Smo is involved in the direct oxidation of spermine in mammalian cells (Cervelli et al. 2004; Bianchi et al. 2005; Murray-Stewart et al. 2008). An acetylspermidine oxidase was overexpressed in *Candida boidinii* and localized at the peroxisomes (Nishikawa et al. 2000). Similarly, experimental evidence has demonstrated that monoamine and Paos colocalize in the peroxisomes of plant, fungal, and mammalian cells (depicted in Figures 2.2 through 2.4). This localization is expected since these proteins present peroxisome targeting signals (Nishikawa et al. 2000; Wu, Yankovskaya, and McIntire 2003; Valdes-Santiago et al. 2010).

Regarding the cellular localization of ornithine decarboxylase (Odc, responsible for putrescine biosynthesis), it was confined to the rough endoplasmic reticulum and to a lesser extent to mitochondrial membranes and nuclei in tissue samples from the cerebral cortex, striatum, and hippocampus of Mongolian gerbils (Anehus et al. 1984; Muller et al. 1991). Schipper and Verhofstad (2002) reviewed the distribution patterns of Odc in the cells of tissues obtained by several techniques. The results showed that Odc was found exclusively in the cytoplasm and nuclei and that Odc localization was dependent on factors such as the cell type and physiological status, and they pointed out the importance of the intracellular translocation for Odc regulation. In *Arabidopsis thaliana*, Odc activity was found in nuclei and chloroplast

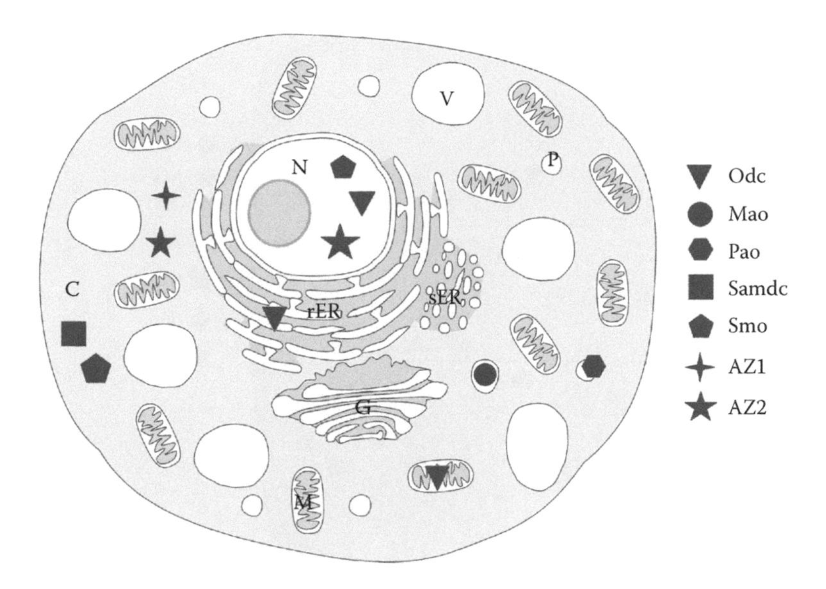

Figure 2.2 Subcellular localization of the enzymes involved in polyamine metabolism in an animal cell. The enzymes included are only those experimentally confirmed. Subcellular components are represented with capital letters as follows: C, cytoplasm; G, Golgi apparatus; M, mitochondria; N, nucleus; P, peroxisomes; rER, rough endoplasmic reticulum; sER, smooth endoplasmic reticulum; V, vacuoles.

(Tassoni, Fornale, and Bagni 2003), whereas there is evidence that at least in the rat kidney, the transcripts of Odc and Ssat are differentially localized, which could indicate that polyamine biosynthesis and degradation may occur at separate cellular compartments (Bettuzzi et al. 1995). In this regard, the differential distribution and cellular localization of polyamines and the enzymes involved in the biosynthesis and retro-conversion of polyamines have been reported in different types of animal tissues and cells (reviewed by Bernstein and Muller 1999).

The so-called "antizyme" is a negative regulator of polyamine synthesis and uptake, and among mammals, there are three antizyme isoforms: AZ1, AZ2, and AZ3. By fluorescence microscopy, it was possible to observe that AZ1 was predominantly localized in the cytoplasm of Chinese hamster ovary cells (Murai, Murakami, and Matsufuji 2003), whereas AZ2 displayed a nuclear dominant distribution and a diffuse distribution in both cytoplasm and nucleus (depicted in Figure 2.2) (Murai et al. 2009).

By means of cytochemical methods, it has been possible to determine the localization of spermidine and spermine *in situ* (Hougaard 1992). Fluorescamine, *o*-phthalaldehyde-mercaptoethanol, and methoxydiphenylfuranone are able to detect primary amino groups, although the lack of specificity, owing to the presence of other substances such as amino acids, amines, amino sugars, and amino phospholipids, may affect the results. A different procedure is the use of radioactive detection methods. By use of these, a possible correlation between the concentration of

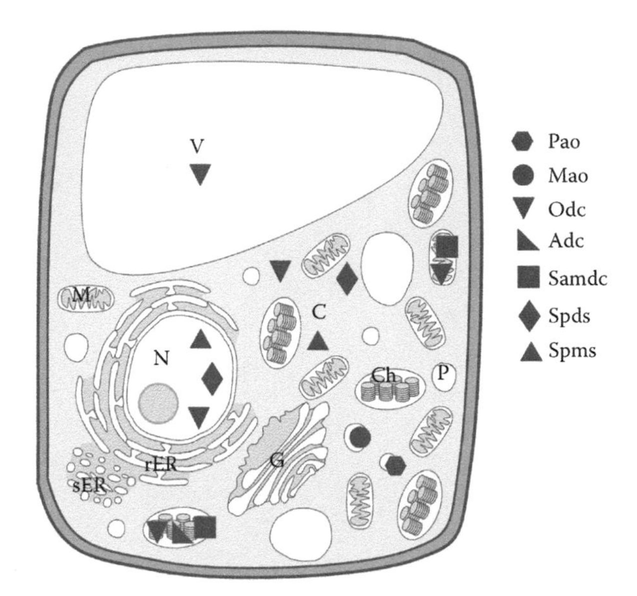

Figure 2.3 Subcellular localization of the enzymes involved in polyamine metabolism in a plant cell. The enzymes included are only those experimentally confirmed. Subcellular components are described with capital letters as follows: C, cytoplasm; Ch, chloroplast; G, Golgi apparatus; M, mitochondria; N, nucleus; P, peroxisomes; rER, rough endoplasmic reticulum; sER, smooth endoplasmic reticulum; V, vacuoles.

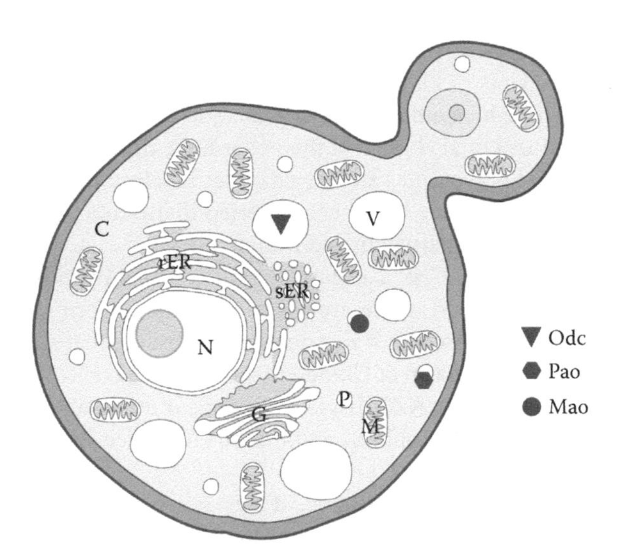

Figure 2.4 Subcellular localization of the enzymes involved in polyamine metabolism in a fungal cell. The enzymes included are only those experimentally confirmed. Subcellular components are described with capital letters as follows: C, cytoplasm; G, Golgi apparatus; M, mitochondria; N, nucleus; P, peroxisomes; rER, rough endoplasmic reticulum; sER, smooth endoplasmic reticulum; V, vacuoles.

spermidine and spermine in tissues such as rat pancreas, rapidly proliferating cells, and white blood cells has been reported (Larsson, Morch-Jorgensen, and Hougaard 1982; Hougaard 1992). In plant cells (*A. thaliana* and *Nicotiana benthamiana*), by means of immunohistological methods using polyclonal antibodies and translational fusions to the green fluorescent protein, it has been shown that aminopropyl transferases (Spds and Spms) displayed a dual subcellular localization both in the cytosol and preferentially inside the nucleus as heterodimer complexes; in contrast, localization of Spms was not clear since a diffused fluorescent signal was found, suggesting that its localization could depend on the cellular context (Belda-Palazon et al. 2012).

The development of polyamine-directed antibodies also has given information regarding the cellular localization of polyamines. Thus, immunoelectron microscopic studies in neurons of the lateral reticular nucleus of rat medulla oblongata showed that spermidine and spermine were predominantly located on free and attached ribosomes to the rough endoplasmic reticulum and in the cytoplasm, but not in the nuclei. These results suggested that polyamines were involved in the translation processes of protein biosynthesis (Fujiwara et al. 1998). Other authors reported that spermine was found in nuclei of differentiated KM-3 cell line, suggesting a possible role of spermine in gene activation (Anehus et al. 1984). Using an antiserum against recombinant human Samdc, both the enzymatically active and inactive forms were located in the cytoplasm of different mouse tissues (Gritli-Linde, Holm, and Linde 1995).

When fluorescent labeling methods were used in human lung carcinoma cells to map the intracellular distribution of polyamine analogues, it was observed that the vast majority of fluorescence was localized in the cytoplasm, and not in the nucleus (Cullis et al. 1999). From these results, it is not possible to conclude if this process was happening *in vivo*, since polyamine analogues appear not to have access to the nucleus.

An immunocytochemical analysis for the localization of polyamines in retinal pigment epithelial and intestinal epithelial cells revealed a population of polyamine immunoreactive vesicles, generally located near the plasma membrane and at a distance from the nucleus. But in another stage of the same cells, polyamine-labeled vesicles were found near the nucleus. The authors hypothesized that after polyamines accumulate in the nuclei (being involved in stabilizing or condensing chromatin), they are removed from the nucleus and transferred to the plasma membrane in vesicles that could act indirectly by influencing the accumulation/storage/release of factors that, in turn, have direct effects on the cell (Johnson et al. 2004).

There are other several reports of staining of vesicles with polyamine antibodies, and it is interesting to notice that polyamine vesicles are confined to cell types that secrete certain growth factors in secretory granules, which could indicate a role of polyamines in the accumulation of vesicle contents and also suggest that polyamines may be released to act alone or in concert with other released factors. Soulet et al. (2004) reported the accumulation of labeled spermidine in vesicles that strongly colocalized with acid vesicles of the late endocytic compartment and the trans-Golgi in cell lines of Chinese hamster ovary. The authors suggested that polyamine transport could be initiated by a membrane carrier and followed by sequestration of the substrate into vesicles via an active mechanism that required an outwardly directed

H^+ gradient. However, there is no evidence that these vesicles were truly secretory, although the results suggested that polyamines were packaged in vesicles and then delivered to the plasma membrane for their release either extracellular or intracellular (Hougaard and Larsson 1986; Hougaard, Del Castillo, and Larsson 1988). Finally, it may be indicated that lysosomes have been suggested as possible storage sites for polyamines in different types of cells (Sindhu and Cohen 1984; Pistocchi et al. 1988; Colombo, Cerana, and Bagni 1992). Although there are no reports about the presence of the polyamine biosynthetic pathway in mitochondria, spermidine and spermine have been detected in the mitochondrial matrix. The presence of a specific mitochondrial polyamine transporter that has also been identified (Toninello, Salvi, and Mondovi 2004) may explain their localization in the organelle.

Regarding fungi, experiments with [^{14}C] ornithine in *Neurospora crassa* showed that spermidine was localized in vacuoles, where it was sequestered and immobilized, since new molecules of spermidine were formed when ornithine was added to the medium, while labeled spermidine was not accessible to be used as substrate for Spms (Paulus, Cramer, and Davis 1983). Odc localization in different compartments has been also observed in *Mucor rouxii*. Using dimethylaminoethyl-dextran (DEAE-dextran) to permeabilize the cells preserving the integrity of internal membranes, Odc activity was measured, and a lower Odc activity compared with cells treated with toluene/ethanol, which destroys intracellular membrane integrity, was observed. These results indicated that Odc was confined in organelles impermeable to ornithine, thus avoiding contact with the external substrate (Martinez-Pacheco and Ruiz-Herrera 1993). These authors also reported that the differential sensitivity to Odc inhibitors of whole or permeabilized cells suggested the existence of at least three different intracellular Odc reservoirs. Something similar happened with *Yarrowia lipolytica* analyzed under conditions that induced the yeast to mycelium dimorphic transition. It was observed that inhibition of Odc stopped the dimorphic transition, but only at early periods, suggesting (although different explanations were possible) that the inhibitor was able to access the enzyme only at the initial period of the transition, whereas at later periods, Odc was not sensitive to the inhibitor, possibly because of Odc compartmentalization, as was demonstrated by comparative assays between intact and permeabilized cells (Guevara-Olvera, Calvo-Mendez, and Ruiz-Herrera 1993). In agreement with the existence of polyamine and synthetic enzyme reservoirs in fungi, as indicated previously, polyamine vesicles have been found also in animal cells, and lysosomes seem to have an important role in the accumulation of these molecules (Sindhu and Cohen 1984; Hougaard and Larsson 1986; Hougaard, Del Castillo, and Larsson 1988; Pistocchi et al. 1988; Colombo, Cerana, and Bagni 1992). In plant cells, Adc, Odc, and Samdc activities were detected in several subcellular fractions, probably also compartmentalized (Torrigiani et al. 1986). The biological relevance of sequestration of polyamines and the enzymes implicated in their metabolism could be related with the translocation of polyamines through the cell, or even between tissues or organs (Kakkar, Rai, and Nagar 1997; Ohe et al. 2005). The ability of polyamines to move freely would allow the rapid response to polyamine requirements, and it would help to explain the complex regulation of polyamine levels; e.g., tobacco transgenic plants overexpressing Odc did not present a significant elevation in spermidine and spermine levels, which could be explained by sequestration of polyamines in vacuoles (DeScenzo and Minocha 1993; Bastola and Minocha 1995; Mayer and Michael 2003).

There are two pathways by which it has been reported that ornithine, the precursor of putrescine, is synthesized in some fungi. In *N. crassa* and *S. cerevisiae*, ornithine is generated from two sources: from glutamate in mitochondria (where the enzymes involved in their synthesis are present in the matrix) or from cytosol by hydrolysis of arginine with arginase. It has been reported that in *S. cerevisiae*, in the presence of both arginine and ornithine, arginase binds reversibly to ornithine carbamoyltransferase (enzyme involved in ornithine synthesis from glutamante), a process that leads to the inhibition of its activity, a peculiar type of control, denominated epiarginasic regulation. This fact is relevant in terms of the cellular localization of the enzymes, and it requires that both enzymes are present in the same cellular compartment, in such way that fungi whose ornithine carbamoyltransferase is localized at the mitochondria do not display epiarginasic regulation (Jauniaux, Urrestarazu, and Wiame 1978).

Using [^{3}H]DFMO (an inhibitor of Odc that binds covalently to the enzyme) to radioactively label Odc, and its subsequent separation by sodium dodecyl sulfate polyacrylamide gel electrophoresis (SDS-PAGE) and fluorography, Odc was found in the cytosol but not in the chloroplast of the unicellular green alga *Chlamydomonas reinhardtii* (Voigt, Deinert, and Bohley 1999). Using the same procedure of specific labeling of Odc with [^{3}H]DFMO, used together with electron microscopic autoradiography, the localization of the enzyme in tomato tissues was analyzed. In this study, radioactivity (i.e., Odc) was found in vacuoles, nucleus, and the cytoplasm (Slocum 1991). In opposition, the activities of Odc, Samdc, and Adc were all observed in chloroplasts of *Pinus radiata* (Torrigiani et al. 1986). Agreeing with this observation, Adc has been localized in chloroplasts of *Avena sativa* (Borrell et al. 1995). A schematic representation of the intracellular localization of polyamines and enzymes involved in their synthesis is depicted in Figure 2.3.

CHAPTER 3

Mechanisms of Polyamines Biosynthesis, Their Interconversion and Degradation in Fungi, in Comparison with Animal and Plant Cells

3.1 INTRODUCTION

As already indicated, the most widely distributed polyamines in living organisms are, in order of chemical complexity, putrescine, spermidine, and spermine. It is important to stress here that polyamine metabolism interacts with the metabolism of amino acids that serve as precursors of polyamines and their degradation byproduct. This makes the regulation of polyamine metabolism more complex than expected, considering that it is not an isolated process, but a component in the cell functional web. This chapter not only compiles investigation in the biosynthetic pathways of polyamines in fungi, as well as other eukaryotic organisms, but it also brings into focus the advances in the knowledge of the interrelation of genes implicated in polyamine metabolism. One advantage in the study of fungal cells is that they contain one single copy of genes involved in polyamine biosynthesis, and others such as *Ustilago maydis* do not have spermine, making the analysis of the physiological roles of the individual polyamines simpler. In this respect, in *U. maydis* mutants unable to synthesize putrescine either by direct or back-conversion, only spermidine and not putrescine was found to be effective in sustaining growth and to be susceptible to engage in a dimorphic transition (Valdes-Santiago, Guzman-de-Pena, and Ruiz-Herrera 2010).

3.2 PATHWAYS OF PUTRESCINE BIOSYNTHESIS

Considering the origin of the three most important polyamines, the main route of putrescine biosynthesis in eukaryotes involves the decarboxylation of ornithine, which in turn is derived from arginine metabolism. Nevertheless, other pathways for putrescine biosynthesis have been described in many bacteria (Hanfrey et al. 2011), some fungi (Valdes-Santiago and Ruiz-Herrera 2014), and in plants (Alcazar et al.

2010; Gupta, Dey, and Gupta 2013) (see further discussion). In bacteria and plants, an alternative pathway for putrescine synthesis involves the action of arginine decarboxylase (Adc). In fungi, however, it is accepted that the most important biosynthetic pathway that has been conserved to generate putrescine involves ornithine decarboxylase (Odc). Putrescine then gives rise to spermidine and spermine by the sequential transfer of two isopropyl units derived from methionine. The detailed mechanisms of all these reactions are discussed further in this chapter.

The pathways involved in polyamine biosynthesis in the different groups of organisms are shown in Figure 3.1. Further in the chapter, we discuss the mechanism involved in the synthesis of each one of the polyamines more common in all organisms: putrescine, spermidine, and spermine.

The synthesis of putrescine in most eukaryotic organisms occurs through decarboxylation of ornithine, which is derived from arginine, and most fungi use this pathway for the synthesis of the polyamine (Tabor and Tabor 1984; Walters 1995). The process of ornithine decarboxylation is catalyzed by Odc (E.C. 4.1.1.17) (see Tabor and Tabor 1984), an enzyme whose control is important for the regulation of polyamine biosynthesis, in general, since it is a rate-limiting enzyme. The amino acid sequences of fungal Odcs may be variable, although their functional structure is highly conserved (Torres-Guzman et al. 1996). The molecular weight of the enzyme is not constant and varies. For example, *Saccharomyces cerevisiae* Odc has an relative molecular mass (Mr) of 86 kDa, whereas *Physarum polycephalum* enzyme has an Mr of 80 kDa, and the Mr of *Neurospora crassa* Odc is 110 kDa. Other Odcs are smaller; thus, the enzyme from *U. maydis* has an Mr of 50.6 kDA, and *Yarrowia lipolytica* Odc contained 449 amino acids with a molecular mass of 49.18 kDa; mouse Odc contained 461 amino acids, giving an Mr of 51 kDa (Tyagi, Tabor, and Tabor 1981; Barnett and Kazarinoff 1984; DiGangi, Seyfzadeh, and Davis 1987; Grens et al. 1989; Yao et al. 1995; Guevara-Olvera, Xoconostle-Cazares, and Ruiz-Herrera 1997; Jimenez-Bremont, Ruiz-Herrera, and Dominguez 2001; Coleman et al. 2004), whereas Odc from the red seaweed *Grateloupia imbricata* presented a molecular weight of 45.9 kDa, similar to the Odc from the nematode *Panagrellus redivivus* with 47.1 kDa (von Besser et al. 1995; García-Jiménez et al. 2009). Odc from tobacco cells and tomato ovaries showed to be similar to the one from mammalian cells, with a large Mr for both of them of about 107 kDa.

Biochemical studies of Odc in higher eukaryotes are widely detailed. The enzymatic activity of Odc depends on the formation of a head-to-tail homodimer (Figure 3.2) (Coleman et al. 1994, 2004; Myers et al. 2001). As a cofactor, Odc uses pyridoxal L-phosphate (Poulin et al. 1992; Myers et al. 2001). Mutation at its *C*-terminal revealed its importance to Odc stability, since it is responsible for its rapid turnover (Ghoda et al. 1989; Li and Coffino 1992). Odc also contains a PEST region (region rich in proline, glutamic acid, aspartic acid, serine, and threonine) at its *C*-terminal region, typical of proteins with short half-lives (Rechsteiner and Rogers 1996); thus, Odc half-life is very short in mammalian cells and it is degraded by the 26S proteasome (Murakami et al. 1992; Elias et al. 1995; Coffino 2001a). As expected, ODC uses pyridoxal phosphate as a cofactor, with reported variable values of michaelis constant (Km) being of 0.14 mM in some plants, while

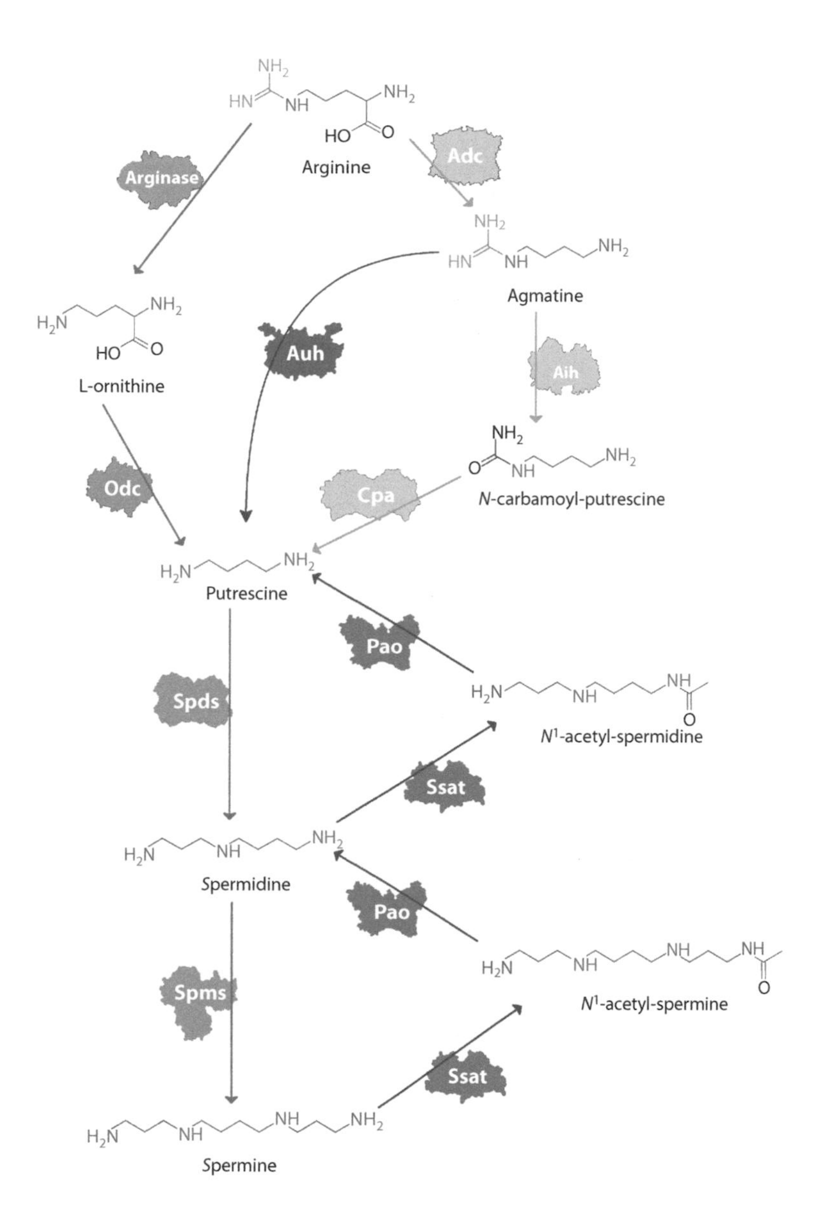

Figure 3.1 **(See color insert.)** Polyamine biosynthetic and retroconversion pathways. Putrescine structure and/or 1,4-diamino-butanoyl groups are in red; amino-propyl groups are in green; the carbamoyl group is in magenta; imino groups are in blue; and carboxyl groups are in black. Enzymes are represented by cartoons, with white letters indicating their respective names. Enzymes involved in the biosynthetic polyamine pathway are in red; enzymes of the alternative putrescine biosynthesis starting with the Adc (arginine decarboxylase) are in green; those enzymes implicated on the back conversion pathway are in magenta; and the agmatine ureohydrolase (Auh) is in blue. Aih, agmatine iminohydrolase; Cpa, *N*-carbamoyl-putrescine amidohydrolase; Odc, ornithine decarboxylase; Pao, polyamine oxidase; Spds, spermidine synthase; Spms, spermine synthase; Ssat, spermine, spermidine N^1-acetyltrasferase.

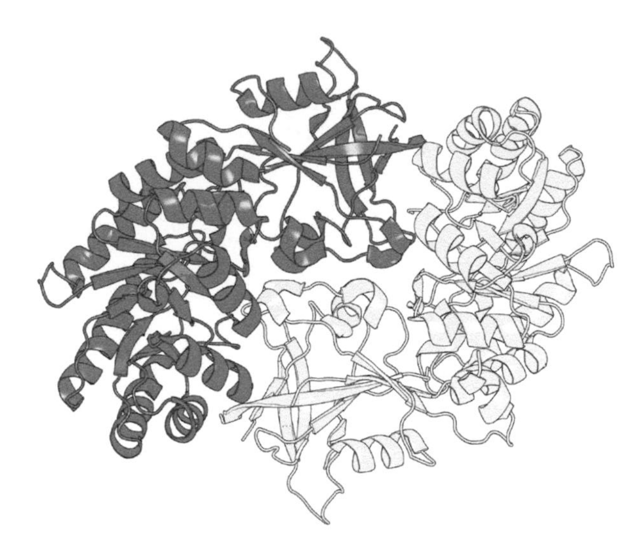

Figure 3.2 The dimeric structure of ornithine decarboxylase (Odc) from *Trypanosoma brucei*, 1F3T (Jackson, L.K., Brooks, H.B., Osterman, A.L., Goldsmith, E.J., Phillips, M.A., *Biochem*, 11247–11257, 2000). The structure was obtained from the RCSB Protein Data Bank, http://www.rcsb.org (Berman, H.M., Westbrook, J., Feng, Z., Gilliland, G., Bhat, T.N., Weissig, H., Shindyalov, I.N., Bourne, P.E., *Nucleic Acids Res.*, 28, 235–242, 2000), and the figure was created with Pymol (DeLano, W.L., *The PyMOL Molecular Graphics System*, DeLano Scientific LLC, San Carlos, CA, 2002; http://www.pymol.org). Each monomer is highlighted with different levels in the gray scale.

in mammalian Odcs, Km has been reported to be 0.09 mM (Pegg and McGill 1979; Heimer and Mizrahi 1982; Seely, Pösö, and Pegg 1982; Osterman et al. 1994; Coleman et al. 2004). Datura Odc is also similar to the other eukaryotic Odcs at the amino acid sequence level; however, it did not have the 3′ extension implied in its extremely fast turnover (Michael et al. 1996). Odc from *Trypanosoma brucei* and *Leishmania donovani* presented high stability and slow turnover, apparently also because the lack of the characteristic sequences at the *C*-terminal region of the mammalian cells enzymes (Ghoda et al. 1990). Odc from other organisms such as *P. redivivus* also has no PEST sequences (von Besser et al. 1995). With regard to fungal enzymes, two PEST regions have been identified in the enzymes from *Y. lipolytica*, *U. maydis*, and *N. crassa* (Williams et al. 1992; Guevara-Olvera, Xoconostle-Cazares, and Ruiz-Herrera 1997; Jimenez-Bremont, Ruiz-Herrera, and Dominguez 2001), whereas only a single putative PEST sequence was described to be present in the Odc from *Candida albicans* (López et al. 1997).

ODC genes and their transcripts have been analyzed in several mammalian cells (Katz and Kahana 1988; Wen, Huang, and Blackshear 1989; Moshier et al. 1990; van Steeg et al. 1990; Yao et al. 1998). In opposition to fungi, mammals contained more than one *ODC* gene, although in mouse, only one of the two *ODC* genes was found to be active (Cox et al. 1988). In the apple genome, a putative *ODC* gene has also been identified, but Odc activity was found only in the apple fruit and calli, but not in other apple tissues (Biasi, Costa, and Bagni 1991; Hao et al. 2005).

In contrast to most fungi, there is another general pathway for the synthesis of putrescine used mainly by plants and bacteria. This pathway involved the action of Adc (EC 4.1.1.19) and in fungi has been reported only in a few species. Accordingly, *Gigaspora rosea*, *Colletotrichum truncatum*, *Ophiostoma ulmi*, and *Laccaria proxima* are some examples of fungi that seem to possess Adc pathway for the synthesis of putrescine (Biondi, Polgrossi, and Bagni 1993; Gamarnik, Frydman, and Barreto 1994; Zarb and Walters 1994; Sannazzaro et al. 2004) (for comparison between plants and fungi polyamine metabolism, see Valdes-Santiago et al. 2012a). It has been described that the Adcs reported are proteins of about 728 amino acids (78 kDa; Hao et al. 2005), and the crystal structure of Adc has been revealed in *Escherichia coli* and *Methanococcus jannaschii*. It has been determined that Adc from *E. coli* is active only under acid conditions. Under these conditions, there occurs the formation of a decamer, as is shown in Figure 3.3, and this is a mechanism of modulating Adc activity by external pH (Tolbert et al. 2003; Andréll et al. 2009). In *Arabidopsis thaliana*, the Adc active form has a head-to-tail homodimer arrangement (Hanfrey et al. 2001). This enzyme carries out the decarboxylation of L-arginine to generate agmatine. In turn, agmatine iminohydrolase, also known as agmatine deiminase (Aih, EC 3.5.3.12), is found in

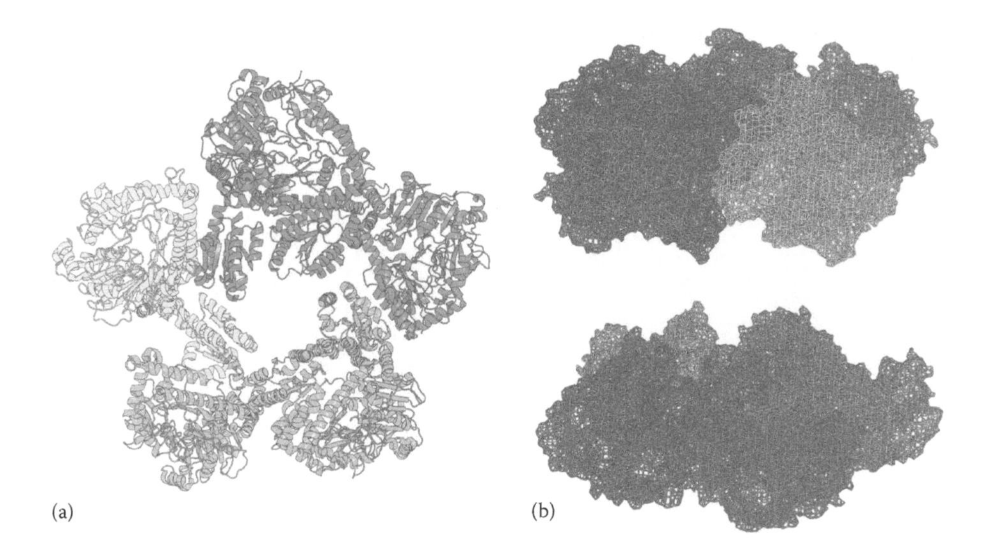

Figure 3.3 **(See color insert.)** Structure of the acid-induced arginine decarboxylase (Adc) from *Escherichia coli*, 2VYC (Andréll, J., Hicks, M.G., Palmer, T., Carpenter, E.P., Iwata, S., Maher, M.J., *Biochem* 48, 3915–3927, 2009) created with Pymol (DeLano, W.L., *The PyMOL Molecular Graphics System*, DeLano Scientific LLC, San Carlos, CA, 2002; http://www.pymol.org). (a) Pentameric ring of Adc. (b) Full decameric structure of Adc represented by a mesh-surface visualization. Each monomer is highlighted with a different color.

A. thaliana as a dimer (Figure 3.4) (Janowitz, Kneifel, and Piotrowski 2003). Aih produces *N*-carbamoyl putrescine, and this in turn is converted to putrescine by the action of *N*-carbamoyl-putrescine amidohydrolase (EC 3.5.1.53) (Figure 3.1, highlighted with green arrows). Plant Odcs have been characterized from different sources, but it is important to notice that the Adc pathway prevails, especially in higher plants. It had been previously considered that *A. thaliana* had no *ODC* gene and that polyamine biosynthesis was achieved only through the Adc activity; however, an Odc with low homology to that present in other organisms has been further identified in its genome (Hanfrey et al. 2001; Tassoni, Fornale, and Bagni 2003). Plants use mostly the Adc pathway in response to different types of stresses, whereas the Odc pathway was reported to be related with plant defense against pathogens in *Capsicum annuum* L, and in tobacco cell cultures, it was described that Odc was induced by methyl jasmonate (Slocum, Kaur-Sawhney, and Galston 1984; Imanishi et al. 1998; Yoo et al. 2004).

In another pathway described in *E. coli*, putrescine is obtained from agmatine directly by the action of agmatine ureohydrolase (Auh), known also as agmatinase, whose active structure is in the form of a trimer (Figure 3.5) (Szumanski and Boyle 1990). The latter pathway is highlighted in blue in Figure 3.1 (Lu et al. 2002; Ahn et al. 2004). The same pathway has been reported to be used by many plants (together with the Adc pathway), by some bacteria, and by some mammalian cells such as neurons and liver cells, but in fungi, it seems to be absent (Nakada and Itoh 2003; Manos et al. 2008; Landete et al. 2010; Suzuki et al. 2013; Wang et al. 2014b).

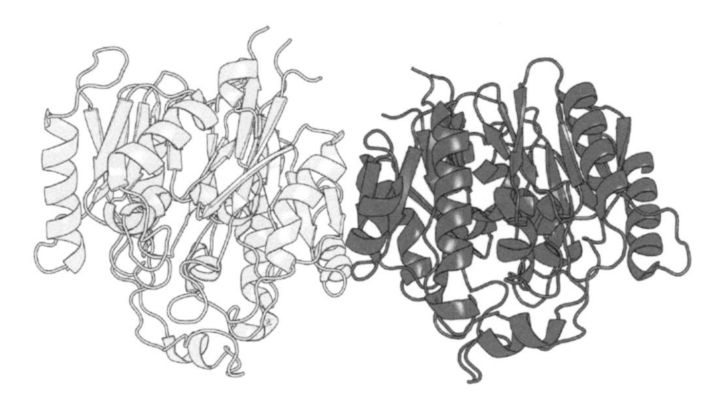

Figure 3.4 Structure of agmatine iminohydrolase (Aih) from *Arabidopsis thaliana*, 2Q3U (Levin, E.J., Kondriashov, D.A., Wesenberg, G.E., Phillips, G.N., *Structure* 15, 1040–1052, 2007). The structure was obtained from the RCSB Protein Data Bank, http://www.rcsb.org (Berman, H.M., Westbrook, J., Feng, Z., Gilliland, G., Bhat, T.N., Weissig, H., Shindyalov, I.N., Bourne, P.E., *Nucleic Acids Res.*, 28, 235–242, 2000), and the figure was created with Pymol (DeLano, W.L., *The PyMOL Molecular Graphics System*, DeLano Scientific LLC, San Carlos, CA, 2002; http://www.pymol.org). Each monomer is highlighted with different levels in the gray scale.

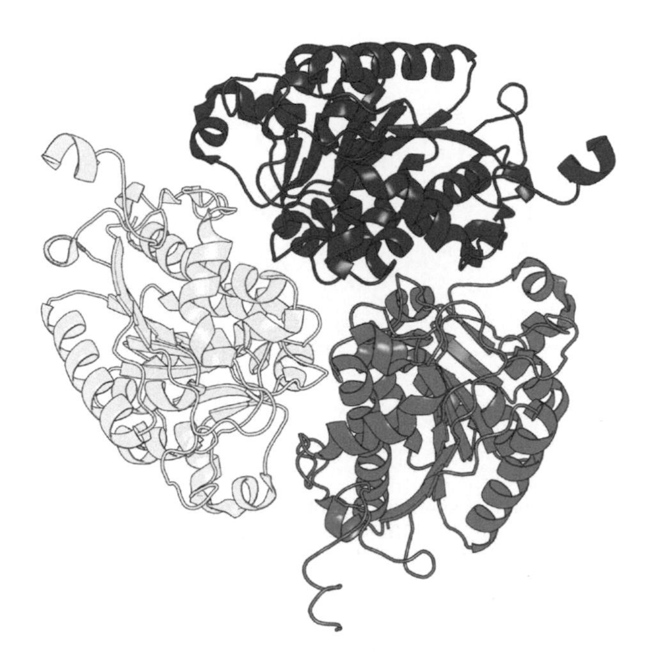

Figure 3.5 Crystal structure of agmatine ureohydrolase (Auh) of *Thermoplasma volcanium*, PDB ID: 3PZL. The structure was obtained from the RCSB Protein Data Bank, http://www.rcsb.org (Berman, H.M., Westbrook, J., Feng, Z., Gilliland, G., Bhat, T.N., Weissig, H., Shindyalov, I.N., Bourne, P.E., *Nucleic Acids Res.*, 28, 235–242, 2000). This figure was created with Pymol (DeLano, W.L., *The PyMOL Molecular Graphics System*, DeLano Scientific LLC, San Carlos, CA, 2002; http://www.pymol.org). Each monomer is highlighted with different levels in the gray scale.

3.3 MECHANISMS FOR THE SYNTHESIS OF SPERMIDINE

The conversion of putrescine to spermidine is catalyzed by the action of an aminopropyltransferase, spermidine synthase (Spds, E.C. 2.5.1.16), whose dimeric structure is shown in Figure 3.6 (H. Wu et al. 2007). Spds transfers an aminopropyl group from decarboxylated *S*-adenosylmethionine (dcSAM) to putrescine. dcSAM is synthesized by *S*-adenosylmethionine (SAM) decarboxylase (Samdc; E.C. 4.1.1.50). Human Samdc exists as a dimer made of two antiparallel eight-stranded β-sheets, as is shown in Figure 3.7 (Ekstrom et al. 1999). Spds produces methylthioadenosine (MTA) as a byproduct (see Figure 3.8). Spds contains an *N*-terminal domain with six-stranded β-sheets, and the *C*-terminal domain consists of seven-stranded β-sheet and nine α-helices (Korolev et al. 2002; Burger et al. 2007). The enzyme from many plants and the parasite *Plasmodium falciparum* present an additional *N*-terminal extension of unknown function (Haider et al. 2005; Dufe et al. 2007). Plants such as pea contain two genes encoding Spds, and *A. thaliana* contains three genes. Their expression patterns seem to be related with the tissue, the plant organs, and even with the stage of development (Alabadi and Carbonell 1999; Hanzawa et al. 2002; Imai

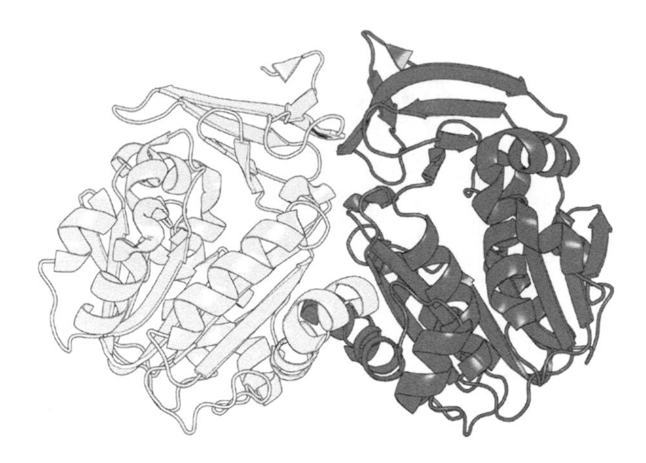

Figure 3.6 Structure of human spermidine synthase (Spds), 2005 (Wu, H., Min, J., Ikeguchi, Y. et al., *Biochem*, 46, 8331–8339, 2007). The structure was obtained from the RCSB Protein Data Bank, http://www.rcsb.org (Berman, H.M., Westbrook, J., Feng, Z., Gilliland, G., Bhat, T.N., Weissig, H., Shindyalov, I.N., Bourne, P.E., *Nucleic Acids Res.*, 28, 235–242, 2000), and the figure was created with Pymol (DeLano, W.L., *The PyMOL Molecular Graphics System*, DeLano Scientific LLC, San Carlos, CA, 2002; http://www.pymol.org). Each monomer is highlighted with different levels in the gray scale.

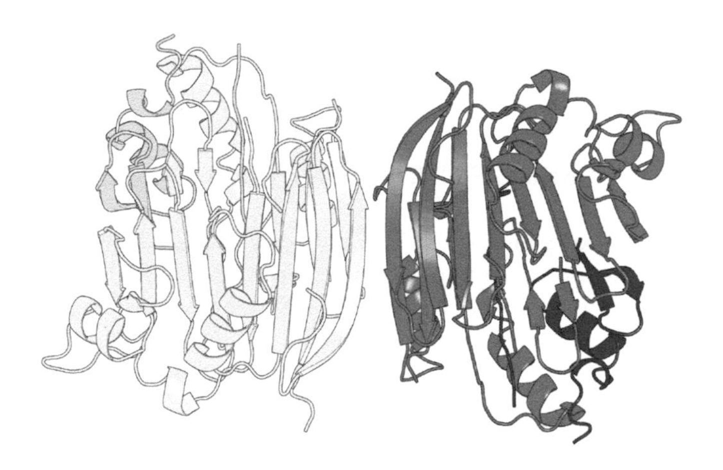

Figure 3.7 Structure of human *S*-adenosylmethionine decarboxylase (Samdc), 1JEN (Ekstrom, J.L., Mathews, I.I., Stanley, B.A., Pegg, A.E., Ealick, S.E., *Structure*, 7, 583–595, 1999). The structure was obtained from the RCSB Protein Data Bank, http://www.rcsb.org (Berman, H.M., Westbrook, J., Feng, Z., Gilliland, G., Bhat, T.N., Weissig, H., Shindyalov, I.N., Bourne, P.E., *Nucleic Acids Res.*, 28, 235–242, 2000), and the figure was created with Pymol (DeLano, W.L., *The PyMOL Molecular Graphics System*, DeLano Scientific LLC, San Carlos, CA, 2002; http://www.pymol.org). Each monomer is highlighted with different levels in the gray scale.

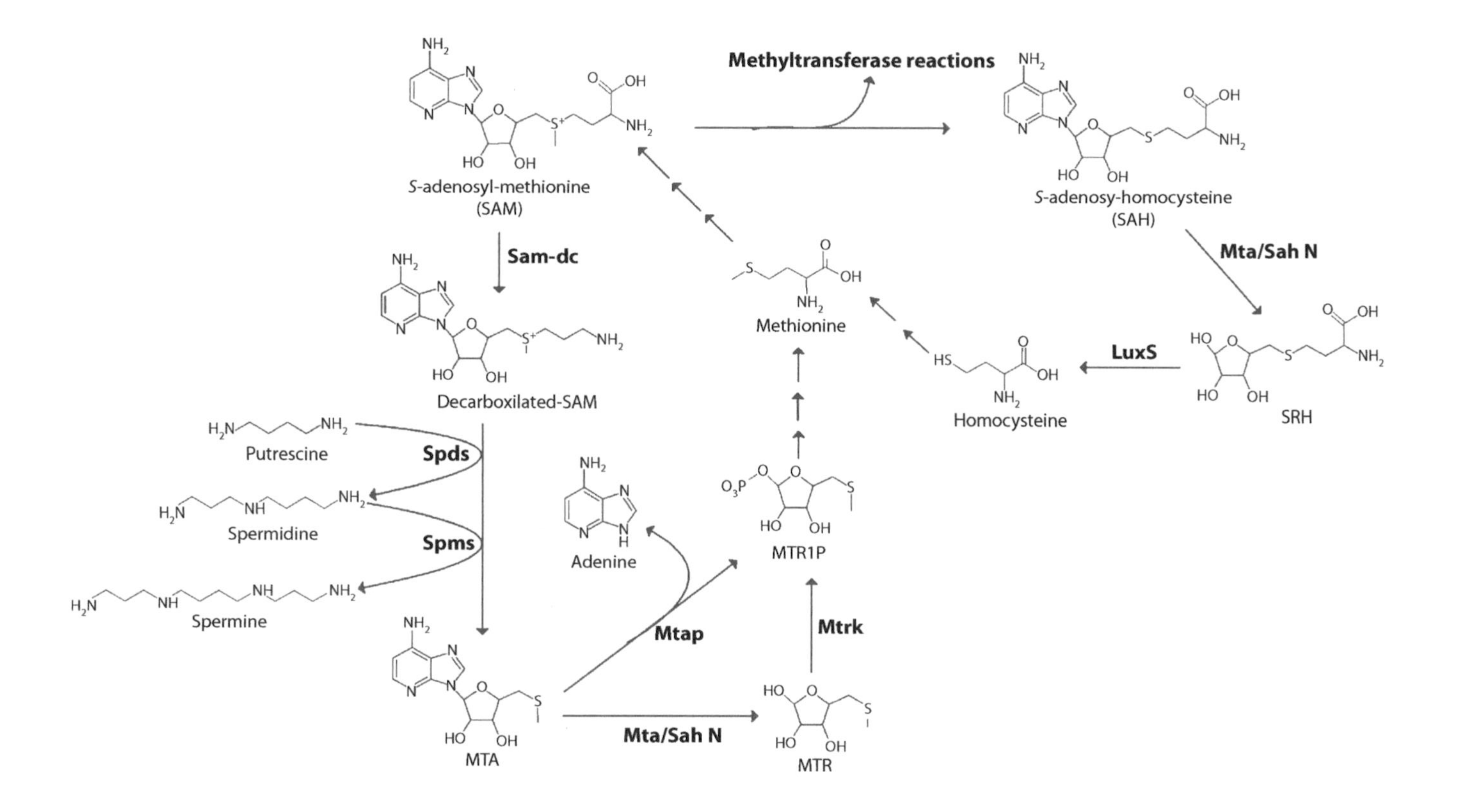

Figure 3.8 Scheme of the methylthioadenosine (MTA) pathway. MTA (left lower end) is generated as a byproduct of polyamine biosynthesis (for details, see text) and metabolized as shown in the scheme. Mtap, MTA phosphorylase; MTR, 5'-methylthioribose; MTPR1P, 5'-methylthioribose 1-phosphate; Mtrk, methylthioribose kinase; Mta/Sah N, 5'-methylthioadenosine/*S*-adenosylhomocysteine (MTA/SAH) nucleosidase; LuxS, *S*-ribosylhomocysteinase; SRH, *S*-ribosylhomocysteine.

et al. 2004a). Interestingly Spds has been immunolocalized in the phloem region of *Solanum tuberosum* (Sichhar and Dräger 2013). In fungi, as in other organisms, spermidine has multiple functions, and one of the most important is the precursor of the eukaryotic initiation factor eIF-5A (see Chapter 6 for details) (Ikeguchi, Bewley, and Pegg 2006; Chattopadhyay, Park, and Tabor 2008). Spds has been mutated in many fungal systems such as the Ascomycota species *Aspergillus nidulans* and *S. cerevisiae* (Hamasaki-Katagiri, Tabor, and Tabor 1997; Jin et al. 2002) and in the Basidiomycota *Cryptococcus neoformans* and *U. maydis*. Interestingly, Spds from *C. neoformans* and *U. maydis* was found to be encoded by a chimeric gene that possesses two regions encoded by the same open reading frame (ORF) (Kingsbury et al. 2004; Valdes-Santiago, Cervantes-Chavez, and Ruiz-Herrera 2009). The first one, corresponding to the *N* terminus, corresponds to Spds, whereas the second one encodes saccharopine dehydrogenase, an enzyme involved in lysine biosynthesis. *C. neoformans* mutants affected in the Spds gene showed an auxotrophy for both lysine and spermidine. Noticeably, our research group demonstrated that this chimeric gene existed in the entire Basidiomycota phylum, being a hallmark of it (Leon-Ramirez et al. 2010).

Samdc is considered the second rate-limiting enzyme (after Odc) in the pathway of polyamine biosynthesis (Stanley, Pegg, and Holm 1989; Pegg et al. 1998). In fungi, and the nematode *Caenorhabditis elegans*, the Samdc encoding gene exists as a single copy in opposition with plants and *L. donovani* that possess at least two copies of the gene (Da'Dara and Walter 1998; Roberts et al. 2002; Valdes-Santiago et al. 2012b). All known Samdcs are unique among enzymes, since they constitute the only group of enzymes that present a covalently bound pyruvoyl group and are synthesized as proenzymes that suffer a cleave at a serine residue (Tabor and Tabor 1987; Pegg et al. 1998; Hillary and Pegg 2003). The particular autocatalytic mechanism of processing and activation of Samdc involves the formation of this pyruvoyl residue from a conserved internal serine that becomes the amino terminal of the α subunit (Stanley, Pegg, and Holm 1989; Kashiwagi et al. 1990a). In fungi, null mutants in the *SAMDC* gene have been isolated from *S. cerevisiae*, *Schizosaccharomyces pombe*, and *U. maydis* (Balasundaram, Tabor, and Tabor 1991; Chattopadhyay, Tabor, and Tabor 2002; Valdes-Santiago et al. 2012b). In mammalian cells, Samdc has a short half-life, similar to Odc (*in vivo* Odc half-life ranked from 10 to 24 minutes, and Samdc, from 22 to 67 minutes), and the degradation of the latter is catalyzed by the 26S proteasome after ubiquitination (Shirahata and Pegg 1985; Berntsson, Alm, and Oredsson 1999; Yerlikaya and Stanley 2004) (the mechanism of Odc degradation is discussed in Chapter 5).

As mentioned previously, all reported Samdcs are synthesized as proenzymes, which are autoactivated by cleavage at an internal serine residue through a serinolysis reaction, to generate two nonidentical subunits, which in turn constitute the active form of the enzyme (Shirahata and Pegg 1986; McCann and Pegg 1992; Bale et al. 2008; Pegg 2009). Putrescine is necessary for the processing and activation of Samdc, at least in yeast, *N. crassa*, and mammalian cells, whereas *E. coli* Samdc is activated by Mg^{2+} (Stevens and Winther 1979; Pegg 1984, 1986; Stanley and Pegg 1991; Hoyt et al. 2000). Interestingly, some fungi, like *Mucor rouxii*, plants, and the protozoan *Tetrahymena pyriformis*, do not require either putrescine or Mg^{2+} to stimulate enzyme activity (Pegg 1984; Calvo-Mendez and Ruiz-Herrera 1991; Xiong et al. 1997). There is evidence in plants that a sequence extension at the *C*-terminal region presented in all known Samdcs of plants is responsible for their activation. Apparently, the residues at the *C*-terminal extension ensure the conformational change required by the proenzymes needed to ensure the processing reaction (Xiong et al. 1997). Yeast Samdc is a peptide with a calculated Mr of 46.214 kDa, which is cleaved to peptides of Mr values of 36.0 kDa (α subunit) and 10.0 kDa (β subunit) (Kashiwagi et al. 1990a), whereas the Samdc proenzyme from *E. coli* has an Mr of 30.4 kDa, and it is cleaved to form two subunits, the first one with an Mr of 12.4 kDa and the second one with an Mr of 18.0 kDa (Tabor and Tabor 1987). In the case of human Samdc, the proenzyme is produced as a 38.3-kDa peptide, which, after autocatalysis and proteolytic cleavage, is transformed into an α subunit of 30.6 kDa Mr and a β subunit of 7.7 kDa (Pegg et al. 1998; Ekstrom et al. 1999).

3.4 MECHANISMS FOR THE SYNTHESIS OF SPERMINE

Putrescine and spermidine seem to be the main polyamines in most fungi (Nickerson and Lane 1977; Pegg and Michael 2010; Valdes-Santiago and Ruiz-Herrera 2014), whereas the more complex polyamine, spermine, is absent in most fungi. The conversion of spermidine to spermine is accomplished, as was already described, by a different aminopropyl transferase, spermine synthase (Spms; E.C. 2.5.1.22).

The crystal structure of human Spms crystal structure has been reported to exist as a dimer of two identical subunits; the dimerization is given by interactions between the *N*-terminal domains (Figure 3.9), and the dimer is the active form (Wu et al. 2008). Spdss, as well as Spms, require dcSAM as the aminopropyl donor required for their synthesis (Pegg 1988; McCann and Pegg 1992; Yerlikaya and Stanley 2004). dcSAM synthesis occurs by the action of SAM decarboxylase, same as what happens for spermidine synthesis. Spms is a ubiquitous enzyme in plants and animals but, as already mentioned, with very limited distribution in fungi (Nickerson, Dunkle, and Van Etten 1977). Unlike Spds, which is present in species from the five fungal phyla, Spms orthologs had been described to be present only in the Saccharomycotina class of the Ascomycota (Pegg and Michael 2010). Nevertheless, our *in vitro* analysis of the distribution of the *SPMS* gene revealed that other Ascomycota groups and even at least a Glomeromycota species contain the gene and therefore probably contain spermine (see data in Chapter 4). The rest of fungi contain only Spds orthologs. Traditionally, spermine has been considered not essential for the normal growth of *S. cerevisiae*, since Spms-less mutants grow as well as the wild-type strain in the absence of exogenous polyamines (Hamasaki-Katagiri et al. 1998; Chattopadhyay, Tabor, and Tabor 2003a). It is also likely that spermine is not essentially required for survival of *Arabidopsis* because mutants affected in the Spms-encoding gene (*SPDS*) were viable, although they do not contain spermine at all (Imai et al. 2004b; Nishimura et al. 2011). Nevertheless, the role of spermine in those organisms able to synthesize it and whether this polyamine can substitute spermidine (which is essential in all the organisms whose role has been studied) still remain elusive (see Chapter 4) (Takahashi and Kakehi 2010). Since Spms is considered not essential for growth, this enzyme has not received as much

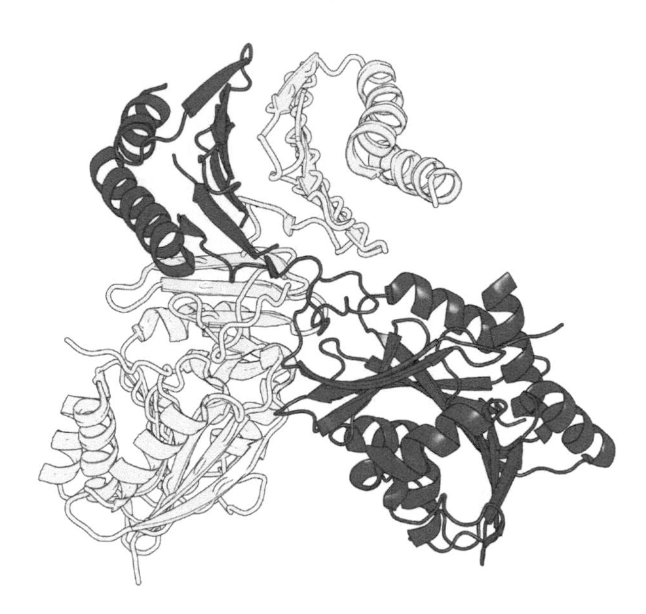

Figure 3.9 Dimeric structure of human spermine synthase (Spms), 3C6M (Wu, H., Min, J., Zeng, H., McCloskey, D.E., Ikeguchi, Y., Loppnau, P., Michael, A.J., Pegg, A.E., Plotnikov, A.N., *J. Biol. Chem.*, 283, 16135–16146, 2008). The structure was obtained from the RCSB Protein Data Bank, http://www.rcsb.org (Berman, H.M., Westbrook, J., Feng, Z., Gilliland, G., Bhat, T.N., Weissig, H., Shindyalov, I.N., Bourne, P.E., *Nucleic Acids Res.*, 28, 235–242, 2000), and the figure was created with Pymol (DeLano, W.L., *The PyMOL Molecular Graphics System*, DeLano Scientific LLC, San Carlos, CA, 2002; http://www.pymol.org). Each monomer is highlighted with different levels in the gray scale.

attention as the rest of the enzymes involved in polyamine biosynthesis. Nevertheless, even when disruption of the gene encoding spermine synthase in mouse embryonic stem cells did not lead to any obvious phenotype at the level of growth rate, morphological alterations, or important modifications of polyamine levels, the mutant showed increased sensitivity to diethyl norspermine, a polyamine analog that, as well as the etoposide, is a cytotoxic anticancer drug, and polyamine synthesis inhibitors such as difluoromethyl ornithine or methylglyoxal bis(guanylhydrazone), inhibitors of Odc and Samdc, respectively (Korhonen et al. 2001). Something similar happened with mouse cell lines that had no Spms activity; they presented enhanced susceptibility to a chloroethylating agent and to shortwave ultraviolet (UV-C) radiation. More impressive is that the absence of spermine affects normal mice development and, in humans, produces the Snyder-Robinson syndrome, an intellectual disability, together with the development of osteoporosis, low muscle mass, and other specific phenotypes (Mackintosh and Pegg 2000; Nilsson, Gritli-Linde, and Heby 2000; de Alencastro et al. 2008). The mechanisms by which spermine is involved in all those physiological processes, as revealed by the phenotype developed by its absence, are not known.

In plants, spermine is considered essential to inflorescence development, and furthermore, plants affected in the Spms gene showed imperfections at the level of elongation of stem internodes and severe affection in xylem specification, probably by retardation on the duration of its differentiation, and in vein thickness. Apparently, polyamines are implicated in polar auxin transport and in the definition of veins (Hanzawa, Takahashi, and Komeda 1997; Hanzawa et al. 2000; Clay and Nelson 2014).

Spds and Spms have some common features; both are homodimeric enzymes with similar native molecular weights between 70 and 80 kDa, and the degradation of both aminopropyltransferases after their ubiquitination occurs by the 26S proteasome (Korolev et al. 2002; Dufe et al. 2005; Ikeguchi, Bewley, and Pegg 2006; H. Wu et al. 2007, 2008). Nevertheless, there are exceptions; thus, Spds from *Thermotoga maritima* is tetrameric (Korolev et al. 2002).

3.5 MECHANISMS OF DEGRADATION AND INTERCONVERSION OF POLYAMINES

Back-conversion and degradation of polyamines involve a well-known process by which spermine is converted into spermidine, which in turn gives rise to putrescine. The process involves acetylation of polyamines that decrease their positive charge so they can be eliminated from the cells either by excretion or by degradation (Seiler 1987). Spermine/spermidine N^1-acetyltrasferase (Ssat; E.C. 2.3.1.57), an enzyme that also exists in a dimeric form (Figure 3.10), catalyzes the acetylation of

Figure 3.10 Dimeric structure of human spermidine/spermine N^1-acetyltransferase (Ssat), 2B5G (Bewley, M.C., Graziano, V., Jiang, J.S., Matz, E., Studier, F.W., Pegg, A.P., Coleman, C.S., Flanagan, J.M., *Proc. Natl. Acad. Sci. USA*, 103, 2063–2068, 2006). The structure was obtained from the RCSB Protein Data Bank, http://www.rcsb.org (Berman, H.M., Westbrook, J., Feng, Z., Gilliland, G., Bhat, T.N., Weissig, H., Shindyalov, I.N., Bourne, P.E., *Nucleic Acids Res.*, 28, 235–242, 2000), and the figure was created with Pymol (DeLano, W.L., *The PyMOL Molecular Graphics System*, DeLano Scientific LLC, San Carlos, CA, 2002; http://www.pymol.org). Each monomer is highlighted with different levels in the gray scale.

polyamines. In turn, N^1-acetyl-spermine and N^1-acetyl-spermidine are substrates to polyamine oxidase (Pao; E.C. 1.5.3.11), yielding spermidine and putrescine, respectively (Seiler 1995; Vujcic et al. 2003). The crystal structure of *S. cerevisiae* Pao has been solved as well as that of Pao from maize and mammalian cells, where it exists as a homodimer (Figure 3.11) (Huang, Liu, and Hao 2005; Fiorillo et al. 2011; Zamora, Melendez, and Galaleldeen 2014).

The mechanism of back-conversion of polyamines is illustrated in Figure 3.1. Ssat is considered a rate-limiting enzyme in the catabolism of spermidine and spermine (Casero and Pegg 1993). As indicated, Ssat products are the corresponding acetyl derivatives of spermidine and spermine, which can have three fates: (1) their transformation into a lower-order polyamine (spermidine from spermine; putrescine from spermidine), (2) their exportation out of the cell, or (3) their degradation (Casero and Pegg 1993). Interestingly, it was observed in *S. cerevisiae* that polyamine acetyltransferase-less mutants (*paa1*), as well as *rad53* mutants (Rad53 is a kinase involved in the regulation of the G_2/M phase checkpoint; Schwartz et al. 2003), curiously have no alterations in their growth rate compared with the wild type, but the double mutants (*paa1/rad53*) were unable to grow at 37°C, which suggested a gene interaction (Liu, Sutton, and Sternglanz 2005). Interestingly, Rad53 hyperphosphorylation is required for filamentous growth of the pathogenic fungus *C. albicans* induced by the DNA-damaging agent methyl methanesulfonate (Wang et al. 2012). Previously, it was reported that the disruption of the *ODC* gene in *S. cerevisiae* (that

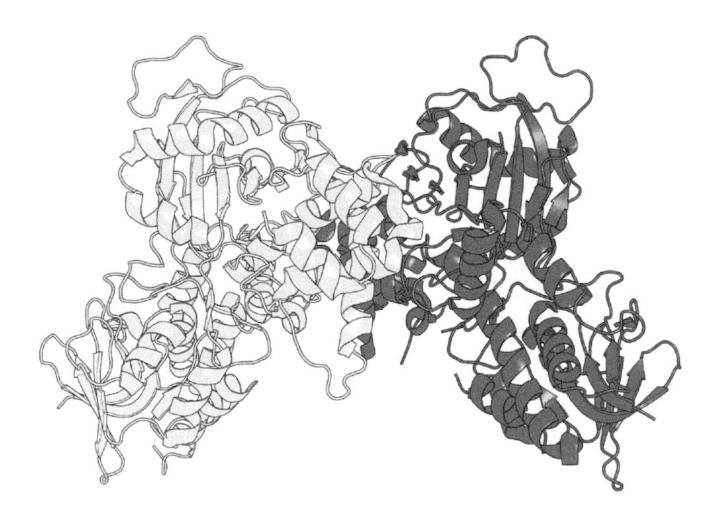

Figure 3.11 Dimeric structure of polyamine oxidase (Pao) from *Saccharomyces cerevisiae*, 1YY5 (Huang, Q., Liu, Q., Hao, Q., *J. Mol. Biol.*, 348, 951–959, 2005). The structure was obtained from the RCSB Protein Data Bank, http://www.rcsb.org (Berman, H.M., Westbrook, J., Feng, Z., Gilliland, G., Bhat, T.N., Weissig, H., Shindyalov, I.N., Bourne, P.E., *Nucleic Acids Res.*, 28, 235–242, 2000), and the figure was created with Pymol (DeLano, W.L., *The PyMOL Molecular Graphics System*, DeLano Scientific LLC, San Carlos, CA, 2002; http://www.pymol.org). Each monomer is highlighted with different levels in the gray scale.

converts the cell into a polyamine auxotroph) restores partially the transcriptional effects caused by mutation of the *GNC5* gene, which encodes a member of the histone acetyltransferase group of enzymes, suggesting a link between polyamines and chromatin modifications (Pollard et al. 1999). It is known that yeast Gcn5 acts in concert with other proteins related with the histone acetyltransferase family, such as Spts, as well as some transcription factors, in a complex named SAGA (Grant et al. 1997; Belotserkovskaya et al. 2000). Additionally, in *S. cerevisiae*, it was found that double mutants *paa1/gcn5*, *paa1/spt8*, and *paa1/spt15* provoked growth defects not displayed by the single mutants (Liu, Sutton, and Sternglanz 2005). Regarding the control of these acetylases, it has been described that factors such as polyamine analogs, hormones, toxic agents, and polyamines proved to be inducers of Paa1 in mammalian cells (Seiler 1987; Pegg 1988; Pegg, Pakala, and Bergeron 1990; reviewed in Casero and Pegg 1993; Pegg 2008). Unfortunately, there is no information in fungi about Paa induction or regulation in fungi, and this issue remains to be solved.

Once polyamines have been acetylated, the flavin-adenine dinucleotide (FAD)-requiring Pao enzyme, a peroxisomal enzyme, oxidizes the acetylated products (Seiler 1995). Oxidation of these N^1-acetyl derivatives transforms spermine into spermidine, and spermidine into putrescine, with the formation of the byproducts acetamidopropanal and H_2O_2 in both cases (Bolkenius and Seiler 1981). The first reported fungal Pao was Fms1 in *S. cerevisiae*, which utilizes N^1-acetylspermine, N^1-acetylspermidine, and N^8-acetylspermidine as its substrates (Landry and Sternglanz 2003; Adachi, Torres, and Fitzpatrick 2010). Disruption of the Pao encoding gene in *U. maydis* led to no obvious apparent phenotype, neither at the level of growth rate, virulence to maize, or response to osmotic or ionic stresses. However, the mutant presented a slightly higher resistance to oxidative stress than the wild type did, and *pao/odc* double mutants were more sensitive to osmotic or ionic stresses and more resistance to the oxidative stress provoked by hydrogen peroxide (Valdes-Santiago, Guzman-de-Pena, and Ruiz-Herrera 2010).

Paos have been classified into two groups. Group I involves those enzymes that act at the terminal amino groups of the substrate, forming ammonia as one of the byproducts; and Group Ib encloses the enzymes that act at the secondary amino groups of the substrates. When the products of these last enzymes are diaminopropanones, they are classified within Group IIa, and when the product is 3-acetaminopropanal, they are classified in Group IIb (Morgan 1998). Most of the reported fungal species contain Paos belonging to Group IIb, as is the case of the Ascomycota *Penicillum* sp., *Aspergillus terreus*, *Hansenula polymorpha*, and *Candida boidinii* (Kobayashi and Horikoshi 1982; Large 1992).

It is important to indicate that catabolic pathway contributes to the regulation of polyamine levels. Two pathways are known to be involved in putrescine degradation (see Figure 3.12), and both of them have 4-aminobutyrate (GABA) as a common intermediate (Seiler and Eichentopf 1975; Schneider and Reitzer 2012). During degradation, putrescine suffers a deamination by the action of 4-aminobutyraldehyde dehydrogenase, whose structure is shown in Figure 3.13. The product, 4-aminobutyraldehyde, is converted into GABA by an aminoaldehyde dehydrogenase that exists as a dimer (Figure 3.14). The degradation of GABA is achieved by 4-aminobutyrate

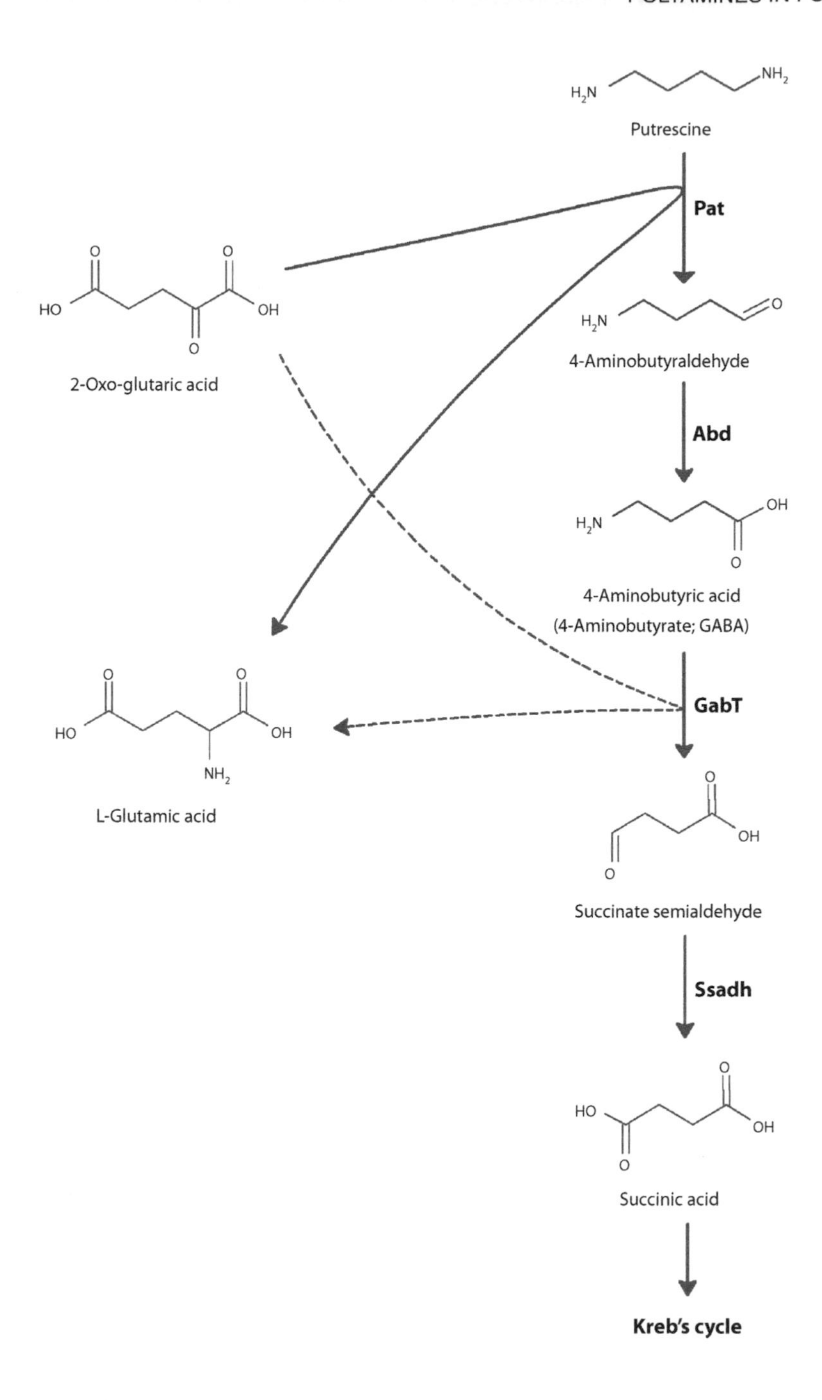

Figure 3.12 Scheme of the putrescine catabolism (for details, see text). Pat, 4-aminobutyraldehyde dehydrogenase; Abd, aminoaldehyde dehydrogenase; GabT, 4-aminobutyrate transaminase; Ssadh, succinate semialdehyde dehydrogenase.

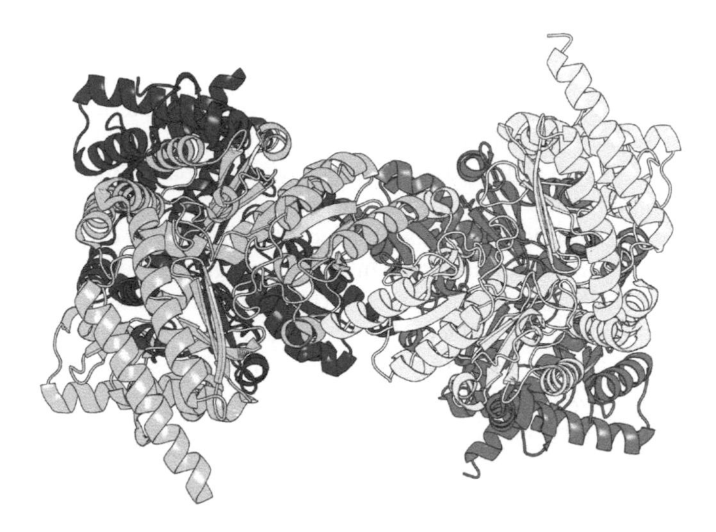

Figure 3.13 Structure of putrescine aminotransferase (Pat) from *Escherichia coli*, 4UOX (Cha, H.J., Jeong, J., Rojviriya, C., Kim, Y., *Plos One*, 9, e113212, 2014). The structure was obtained from the RCSB Protein Data Bank, http://www.rcsb.org (Berman, H.M., Westbrook, J., Feng, Z., Gilliland, G., Bhat, T.N., Weissig, H., Shindyalov, I.N., Bourne, P.E., *Nucleic Acids Res.*, 28, 235–242, 2000), and the figure was created with Pymol (DeLano, W.L., *The PyMOL Molecular Graphics System*, DeLano Scientific LLC, San Carlos, CA, 2002; http://www.pymol.org). Each monomer is highlighted with different levels in the gray scale.

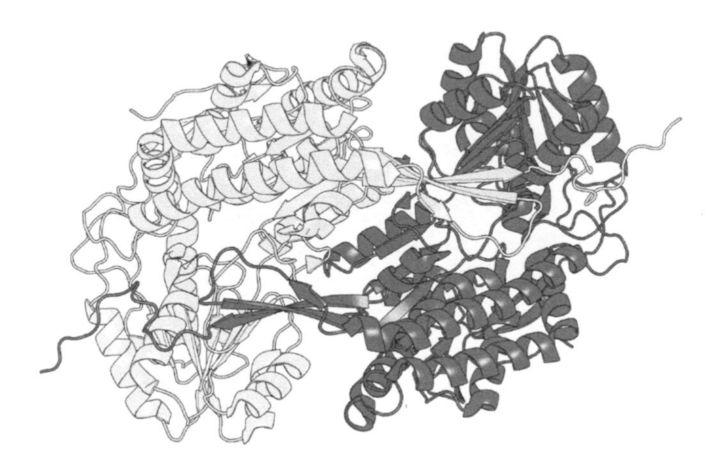

Figure 3.14 Structure of the aminoaldehyde dehydrogenase 2 (Abd) from *Pisum sativum*, 3IWJ (Tylichova, M., Kopecny, D., Morera, S., Briozzo, P., Lenobel, R., Snegaroff, J., Sebela, M., *J. Mol. Biol.*, 396, 870–882, 2010). The structure was obtained from the RCSB Protein Data Bank, http://www.rcsb.org (Berman, H.M., Westbrook, J., Feng, Z., Gilliland, G., Bhat, T.N., Weissig, H., Shindyalov, I.N., Bourne, P.E., *Nucleic Acids Res.*, 28, 235–242, 2000), and the figure was created with Pymol (DeLano, W.L., *The PyMOL Molecular Graphics System*, DeLano Scientific LLC, San Carlos, CA, 2002; http://www.pymol.org). Each monomer is highlighted with different levels in the gray scale.

transaminase with the structure of tetramere (Figure 3.15) to produce succinate semialdehyde, which in turn is transformed to succinic acid by succinate semialdehyde dehydrogenase, a dimeric protein (Figure 3.16). By an alternative mechanism, diamine oxidases degrade putrescine to 4-aminobutyraldehyde/Δ-1-pyrroline, which is transformed to GABA (Jakoby and Fredericks 1959). In plants and bacteria, GABA produced from putrescine catabolism has been implicated in stress response (Shelp et al. 2012; Schneider, Hernandez, and Reitzer 2013). In neuron cell cultures, a relationship between GABA (a well-known inhibitory neurotransmitter) and polyamine metabolism has been established; in fact, a role of putrescine as a neuroprotective molecule has been proposed, since evidences showed that putrescine synthesis is increased and converted to GABA, which in turn activates presynaptic GABA β receptors when *Xenopus laevis* was treated with a convulsant (Seiler, Sarhan, and Roth-Schechter 1981; Bell et al. 2011).

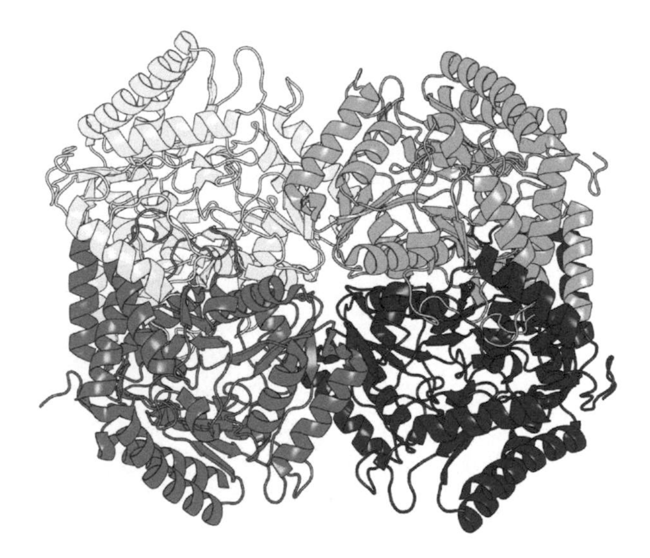

Figure 3.15 Structure of a 4-aminobutyrate aminotransferase (GabT) from *Mycobacterium abscessus*, 4FFC (Baugh, L., Phan, I., Begley, D.W., Clifton, M.C., Armour, B., Dranow, D.M., Taylor, B.M., Muruthi, M.M., Abendroth, K., Fairman, J.W., Fox, D., Dieterich, S.H., Staker, B.L., Gardberg, A.S., Choi, R., Hewitt, S.N., Napuli, A.J., Myers, J., Zhang, Y., Ferrell, M., Mundt, E., Thompkins, K., Tran, N., Lyons-Abbott, S., Abramov, A., Sekar, A., Serbzhinskiy, D., Lorimer, D., Buchko, G.W., Stacy, R., Stewart, L.J., Edwards, T.E., Van Voorhis, W.C., Myler, P.J., *Tuberculosis*, 2014). The structure was obtained from the RCSB Protein Data Bank, http://www.rcsb.org (Berman, H.M., Westbrook, J., Feng, Z., Gilliland, G., Bhat, T.N., Weissig, H., Shindyalov, I.N., Bourne, P.E., *Nucleic Acids Res.*, 28, 235–242, 2000), and the figure was created with Pymol (DeLano, W.L., *The PyMOL Molecular Graphics System*, DeLano Scientific LLC, San Carlos, CA, 2002; http://www.pymol.org). Each monomer is highlighted with different levels in the gray scale.

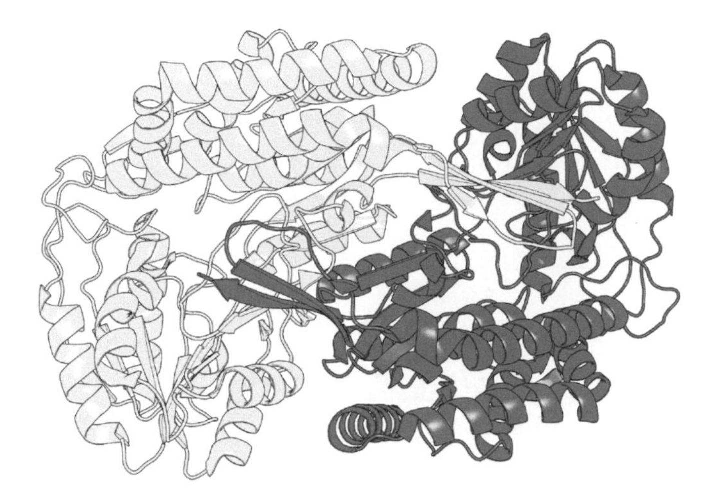

Figure 3.16 Dimeric structure of succinic semialdehyde dehydrogenase (Ssadh) from *Steptococcus pyrogenes*, 4OGD (Jang, E.H., Park, S.A., Chi, Y.M., Lee, K.S., *Mol. Cells*, 37, 719–726, 2014). The structure was obtained from the RCSB Protein Data Bank, http://www.rcsb.org (Berman, H.M., Westbrook, J., Feng, Z., Gilliland, G., Bhat, T.N., Weissig, H., Shindyalov, I.N., Bourne, P.E., *Nucleic Acids Res.*, 28, 235–242, 2000), and the figure was created with Pymol (DeLano, W.L., *The PyMOL Molecular Graphics System*, DeLano Scientific LLC, San Carlos, CA, 2002; http://www.pymol.org). Each monomer is highlighted with different levels in the gray scale.

3.6 ORIGIN AND ROLE OF THE DIFFERENT PRECURSOR MOLECULES INVOLVED IN POLYAMINE BIOSYNTHESIS

As we have indicated, ornithine has an important role as precursor in polyamine biosynthesis. The *de novo* synthesis of L-ornithine has been described in mammalian cells. It is known that glutamine is converted to glutamate semialdehyde, which in turn is converted to ornithine by the action of ornithine aminotransferase. Additionally, L-ornithine originates from L-arginine by the action of arginase (Figure 3.17) during a reaction by which urea is also formed (Jones 1985; Munder 2009; Marini et al. 2012). The synthesis, regulation, and catabolism of arginine and ornithine in *S. cerevisiae* and *N. crassa* have been thoroughly reviewed by Davis (1986). The arginase present in ureotelic species has a main role in nitrogen metabolism in the urea cycle; however, under special circumstances such as pregnancy, arginase is able to supply polyamines (Mora et al. 1965; Weiner et al. 1996). In mammals, arginase I and arginase II display differential expression and different cellular localization, and there are some evidences indicating that these enzymes might be implicated in host immune evasion (Jenkinson, Grody, and Cederbaum 1996; Hai et al. 2014). In parasites like *Leismania mexicana* and *T. brucei*, arginase has been considered a drug target for treatment of their infestations, since its role is to sustain growth by its implication in polyamine biosynthesis (D'Antonio et al. 2013; Hai et al. 2015). In plants, L-arginine is used as nitrogen stock and recycling during its

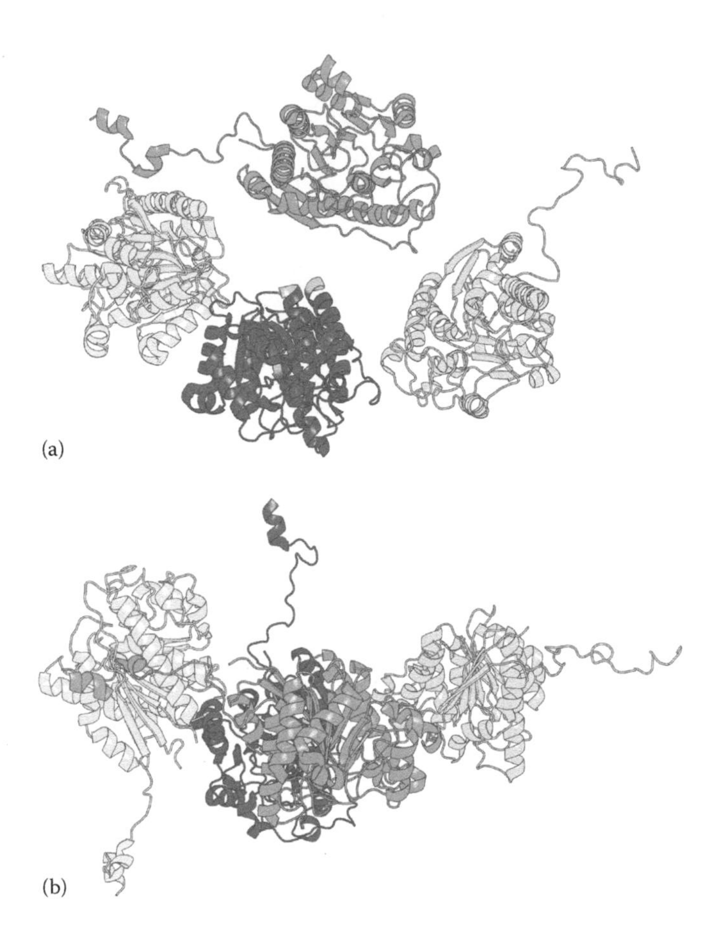

Figure 3.17 The structure of arginase from *Schistosoma mansoni*, 4Q3P (Hai et al. 2014) created with Pymol (DeLano, W.L., *The PyMOL Molecular Graphics System*, DeLano Scientific LLC, San Carlos, CA, 2002; http://www.pymol.org). (a) Upper-view of arginase tetrameric structure. (b) Side-view of (a). Each monomer is highlighted with different levels in the gray scale.

degradation, as well as for the biosynthesis of polyamines by its decarboxylation by Adc (Tiburcio et al. 1985). Also, during seed germination, an increment in arginase activity has been observed (Zonia, Stebbins, and Polacco 1995).

SAM is another precursor required for polyamine biosynthesis. SAM is synthesized from L-methionine by the action of the enzyme SAM synthetase (Tabor and Tabor 1984). Interestingly, SAM is also the substrate for DNA methylases, and in plants, it serves also as a precursor for the synthesis of ethylene, nicotine, and tropane alkaloids (Pandey et al. 2000; Martin-Tanguy 2001).

3.7 THE RELATIONSHIP OF POLYAMINE METABOLISM WITH OTHER CELLULAR PATHWAYS IN DIFFERENT ORGANISMS

When discussing the mechanisms of polyamine biosynthesis, it is necessary to consider that, in addition to be precursor of putrescine, ornithine produces L-proline via ornithine aminotransferase, and L-arginine not only produces L-ornithine, it also is the precursor of nitric oxide, a signal and effector molecule in animals (Morris 2002; Stuehr 2004). Accordingly, the channeling of these products to their different metabolic pathways must be extremely well regulated.

It is possible to say that regulation of polyamine biosynthesis can also be controlled by the production of metabolites related with polyamines such as MTA. MTA is a byproduct of polyamine synthesis since it is generated from dcSAM during the synthesis of spermidine and spermine. MTA is converted into adenine and methylthioribose 1-phosphate by the action of MTA phosphorylase (Mtap). After a series of reactions, both products are converted to adenine nucleotides and methionine, as is shown in Figure 3.8 (Backlund and Smith 1981; Kamatani and Carson 1981; Cone et al. 1982; Savarese et al. 1983; Marchitto and Ferro 1985). *S. cerevisiae* contains one Mtap homolog, and mutants defective in Mtap display a significantly increase in Odc activity, causing the elevation of polyamine concentrations (Subhi et al. 2003). Additionally, it has been suggested that the accumulation of Mta in *S. cerevisiae mtap* mutants inhibits Spds (Chattopadhyay, Tabor, and Tabor 2006b). In the case of mammals, the concentration of MTA regulates Odc and Samdc activities in enzyme-deficient lymphoma cells, provoking an increase in putrescine and dcSAM concentrations (Kubota, Kajander, and Carson 1985).

The interconnection of plant polyamine metabolism with other metabolic pathways, such as generation of nitric oxide and GABA, has also been documented (Stuehr 2004; Alcazar et al. 2010), although the alterations in the phenotype and the mechanisms involved are not well known yet.

In plants, polyamines have been related with abscisic acid (ABA), gibberellins, cytokinins, and nitric oxide. Exogenous applications of these compounds affect polyamine biosynthesis, ABA increases polyamine levels, and its inhibition provokes a reduction in Adc activity; furthermore, under salt stress, there is an elevation in ABA content and a reduction in polyamine levels, while under drought stress, a synchronic accumulation of ABA and an increase in Pao activity have been observed (Lee, Lur, and Chu 1997; Nieves et al. 2001; Cuevas et al. 2008; Shevyakova et al. 2013; Zhang and Huang 2013). All these mechanisms indicate that the interaction of polyamine metabolism with other metabolic reactions of the cells constitute an extremely complicated net and that we must be cautious when attributing to a simple alteration in the synthesis of polyamines a determined phenotypic change in the physiology of the organism.

CHAPTER 4

Polyamine Distribution in the Fungal Kingdom, as Compared with Other Eukaryotic Organisms

4.1 INTRODUCTION

Polyamines have been found in practically every microbial, plant, or animal cell analyzed. In the majority of fungi, putrescine and spermidine can be found, although spermine is absent in most of them. Analysis of the distribution of polyamines in the fungal kingdom has been uncovered experimentally in few species. However, the extensive number of genomes sequenced and the high degree of sequence similarity have allowed the phylogenetic analysis of polyamines in these organisms. In fact, analysis of the polyamine biosynthetic pathway has been considered useful for the study of evolutionary mechanisms (Green et al. 2011).

As already indicated in previous chapters, the pathway for the biosynthesis of the three most common polyamines in fungi and animals is as follows. The first reaction involves the decarboxylation of ornithine, by ornithine decarboxylase (Odc), to form putrescine. Spermidine synthase (Spds) and spermine synthase (Spms) are aminopropyl transferases that produce spermidine and spermine, respectively. Transfer of an aminopropyl group from the donor decarboxylated *S*-adenosylmethionine (SAM) to putrescine by Spds produces spermidine, and the incorporation of a new aminopropyl group to spermidine by Spms gives rise to spermine. *S*-adenosylmethionine decarboxylase (Samdc) is the enzyme that catalyzes the decarboxylation of SAM to synthesize the donor of the aminopropyl groups. As also described before, the pathway of polyamine catabolism involves the assistance of spermidine/spermine-N^1-acetyltransferase (Ssat) and polyamine oxidase (Pao). The first enzyme generates acetyl derivatives of the polyamines, and Pao oxidizes the acetylated products converting spermidine into putrescine and spermine into spermidine (see Chapter 3). This pathway operates for the degradation of polyamines, as well for the transformation of the different polyamines, a mechanism that helps to maintain an adequate proportion of the different polyamines in the cell. Taking advantage of the sequenced fungal genomes, this chapter describes the amino acid sequences of the enzymes involved in several species of fungi, covering most of the fungal kingdom.

This analysis reveals polyamine putative interconnection with pathways such as mitochondrial presequence proteases and peroxisomal fatty acid β-oxidation system.

4.2 POLYAMINES PRESENT IN THE DIFFERENT FUNGAL GROUPS AND OTHER ORGANISMS

Pathways for the biosynthesis of polyamines have been analyzed in a great number of species of animals, plants, and microorganisms. These studies have demonstrated that, in general, prokaryotic and eukaryotic cells synthesize at least putrescine and spermidine (Tabor and Tabor 1984; Bachrach 2010), and some others synthesize also spermine and/or additional polyamines such as cadaverine, norspermidine, and norspermine (Holtta and Pohjanpelto 1983; Kuehn et al. 1990).

It is generally considered that spermine is present in eukaryotes but absent from prokaryotes, but this has not been fully demonstrated, and indeed, it is important to stress that no representative organisms of all the taxonomic groups of eukaryotic organism have been experimentally analyzed to conclude whether they contain or not the three most common polyamines. For example, at the experimental level, in the case of fungi, the activities of Odc and Samdc, as well as the presence or absence of putrescine, spermidine, and spermine, have been confirmed in only a few species, e.g., *Saccharomyces cerevisiae*, *Neurospora crassa*, *Ustilago maydis*, *Mucor hiemalis*, *Rhizopus stolonifer*, *Trichoderma viride*, *Aspergillus nidulans*, *Aspergillus niger*, *Aspergillus oryzae*, *Penicillium expansum*, *Penicillium fumiculosum*, *Penicillium nigricans*, *Alternaria alternata*, *Coccidioides immitis*, *Gigaspora rosea*, *Sclerotinia sclerotiorum*, *Aspergillus fumigatus*, and *Paxillus involutus* are the most widely analyzed fungi (Hart, Winther, and Stevens 1978; Paulus, Kiyono, and Davis 1982; Guevara-Olvera et al. 2000; Valdes-Santiago, Cervantes-Chavez, and Ruiz-Herrera 2009).

Nevertheless, it must be taken into consideration that the conditions of growth, the nature of the culture media, and the presence of different effectors may affect the levels of polyamines or their metabolic enzymes, giving erratic results. For example, some changes in polyamine metabolism occurring by the effect of the environment or during the different developmental stages in fungi have been published for several species. Thus, in the ectomycorrhizal fungus *P. involutus* exposed to zinc, no significant differences were observed in enzyme activities leading to the synthesis of putrescine and spermidine, while spermine concentration changed depending on zinc concentration (Zarb and Walters 1995). In the soil-borne plant pathogen *S. sclerotiorum*, during sclerotial development, the concentration of spermidine and spermine decreased together with the activities of Odc and Samdc, and both activities were minor in sclerotia compared with the mycelial form. This observation showed that some specific morphogenetic stages might be less depending on polyamine or that the control point of the event (when high polyamine levels are required) of the event precedes the phenomenon (Garriz et al. 2008). In the filamentous fungus *A. fumigatus*, the presence of putrescine, spermidine, and spermine was confirmed, and it was reported that the concentration of putrescine and the activity of Odc increased in parallel to biomass production (Walters, Cowley, and McPherson 1997).

Something similar occurs in plants, where the occurrence of putrescine, spermidine, and spermine in plants is vastly known, as well as the existence of polyamine conjugates with caffeic, coumaric, and ferulic acids (Galston and Sawhney 1990). Conjugated polyamines in plants seem to have a role in pathogen resistance and detoxification of phenolic compounds (Martin-Tanguy 2006). It must be recalled that most eukaryotic cells produce putrescine mainly through the action of Odc, while plants and many bacteria use the alternative route for putrescine biosynthesis by the action of arginine decarboxylase (Adc). The presence of this route can be explained in terms of regulation at the level of different tissues or development, as it has been nicely demonstrated in *Nicotiana tabacum*, where Adc activity was found in old hypergeous vascular tissues, while Odc activity was present in hypogeous tissues. Thus, Odc expression was related with early cell divisions and Spds was related with later cell divisions. Also, in general, a spatial and temporal distribution of polyamines was observed, an observation that may explain the advantage of using more than one pathway for polyamine biosynthesis (Paschalidis and Roubelakis-Angelakis 2005).

In plants, differences in polyamine content have been determined at the levels of floral initiation, anthesis, fruit development (in olive and pea), unpollinated parthenocarpic (in tomato), ovary senescence, fruit development (in pea), and even under temperature stressful conditions (in *Arabidopsis*) (Carbonell and Navarro 1989; Fos et al. 2003; Pritsa and Vogiatzis 2004; Todorova et al. 2007). Essentially, differences in polyamine distribution were found depending on the developmental stages, environmental conditions, and photosynthetic activity (Urano et al. 2003). Something similar happens at the transcriptional level of the genes encoding the biosynthetic enzymes. Thus, *Samdc* transcript was found in all tissues of *Phaseolus vulgaris*, while transcripts of *Odc*, *Adc*, and *Spd*s showed tissue specificity (Jimenez-Bremont et al. 2006).

In addition to these three major polyamines, several unusual polyamines have been found in other organisms (Figure 4.1). Among these, we may cite thermospermine, which is apparently widespread in the plant kingdom (Hamana et al. 1994; Oshima 2007; Fuell et al. 2010; Morimoto et al. 2010; Pegg and Michael 2010; Takano, Kakehi, and Takahashi 2012). Norspermidine and norspermine have been detected in some higher plants such as alfalfa (Rodriguez-Garay, Phillips, and Kuehn 1989). Homospermidine and cadaverine were detected in the leaves and roots of aquatic plants, as well as in gramineous plants; aminopropylhomospermidine was detected in the water lily *Nymphaca tetragona* and the lotus *Nelumbo nucifera* and in reduced levels in gramineous seeds (Hamana et al. 1994). The presence of cadaverine and agmatine has been reported in some fungi and plants (Gamarnik and Frydman 1991; Oshima 2007), and Oshima, Moriya, and Terui (2011) described unusual polyamines found in extreme and moderate thermophiles. These unusual polyamines are also found in other microorganisms and lower animals such as lobsters and cockroaches (Stillway and Walle 1977; Oshima, Moriya, and Tervi 2011). On the other hand, the absence of uncommon polyamines such as norspermidine, norspermine, and homospermine was reported for some fungi (Hamana and Matsuzaki 1985). Interestingly,

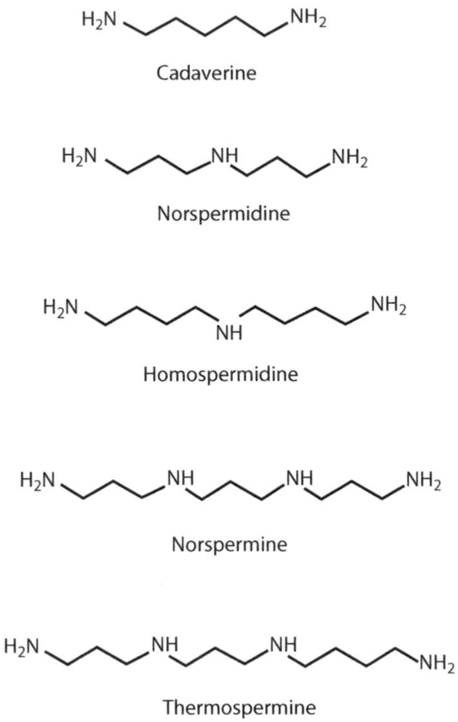

H_2N NH_2

Cadaverine

H_2N NH NH_2

Norspermidine

H_2N NH NH_2

Homospermidine

H_2N NH NH NH_2

Norspermine

H_2N NH NH NH_2

Thermospermine

Figure 4.1 Chemical structure of the most common unusual polyamines present in some bacteria and plants.

there is not much information about the physiological functions of these rare and minor polyamines.

The differences in polyamine distribution and the occurrence of unusual polyamines can be explained either by the elimination of some enzymes of the biosynthetic pathway or by changes and acquisition of a different specificity of existing enzymes during evolution. The absence or presence of certain polyamines, as well as specific features of genes encoding polyamine biosynthetic enzymes, can have taxonomical significance, and even they might be used as phylogenetic and taxonomic markers (Hamana and Matsuzaki 1985; Zherebilo et al. 2001; Hamana et al. 2008; Leon-Ramirez et al. 2010).

4.3 PHYLOGENETIC RELATIONSHIPS OF THE ENZYMES INVOLVED IN POLYAMINE METABOLISM IN FUNGI

It is interesting to notice that differences in the polyamines present in organisms belonging to different kingdoms have been determined by experimental analysis. However, as indicated in Section 4.2, we may be cautious when the absence of a certain polyamine has been reported and consider the possibility that some of these results may possibly be the result of the presence of only small amounts of polyamines that escape to detection, making their analysis at certain stages of growth difficult (Hart, Winter, and Stevens 1978). Analysis of the genomes is an additional tool to conclude if an organism contains specific polyamines, assuming that the genes are operational. Of course, the important limitation of this technique is that the corresponding genomes must have been sequenced. If this has taken place, it is possible to search for homologous genes by means of Blast analysis.

Taking advantage of the homology existing among the polyamine biosynthetic genes, we proceeded to perform *in silico* analyses to determine the distribution of these genes involved in polyamine metabolism in the species of fungi whose genomes have been sequenced. The protein–protein BLAST search showed that Odc, Spd, Samdc, and Pao are widely distributed, while the presence of Spm is limited to only some fungal classes. With these results, we may conclude that the synthesis of both putrescine and spermidine biosynthesis in fungi follows the same pathway used by those species whose analysis has been experimentally made.

The phylum Ascomycota has been divided in three subphyla, Pezizomycotina, Saccharomycotina, and Taphrinomycotina (James et al. 2006; Hibbett et al. 2007). In our analysis, we identified homologous genes in most of the classes of each one of the subphyla, including the following species: *Ascoidea rubenscens*, *Ashbya gossypii*, *Candida arabinofermentans*, *Candida tenuis*, *Debaryomyces hansenii*, *Dekkera bruxellensis*, *Hansenula polymorpha*, *Kluyveromyces lactis*, *Lipomyces starkeyi*, *Pachysolen tannophilus*, *Pichia pastoris*, *S. cerevisiae*, *Spathaspora passalidarum*, *Yarrowia lipolytica*, *Saitoella complicata*, *Schizosaccharomyces pombe*, *Taphrina deformans*, *Aspergillus aculeatus*, *A. niger*, *C. immitis*, *Histoplasma capsulatum*, *Paracoccidioides brasiliensis*, *Penicillium chrysogenum*, *Cenococcum geophilum*, *Cochliobolus victoriae*, *Lentithecium fluviatile*, *Zopfia rhizophila*, *Melanoma pulvis*, *Cladonia grayi*, *Xanthoria parietina*, *Amorphoteca resinae*, *Botrytis resinae*, *S. sclerotiorum*, *Apiospora montaqnei*, *Chaetomium globosum*, *Colletotrichum higginsianum*, *Cryphonectria parasitica*, *Neurospora tetrasperma*, *Phaeoacremonium aleophilum*, and *Podospora anserina*.

The phylum Basidiomycota is phylogenetically divided into three subphyla. These are Pucciniomycotina, Ustilaginomycotina, and Agaricomycotina (James et al. 2006; Hibbett et al. 2007), and we analyzed here the following species: *Agaricus bisporus*, *Armillaria mellea*, *Coprinopsis cinerea*, *Pleurotus ostreatus*, *Rhizoctonia solani*, *Atractiellales* sp., *Cronartium quercuum*, *Melampsora laricis*, *Mixia osmundae*, *Puccinia graminis*, *Rhodotura graminis*, *Sporobolomyces roseus*, *Exobasidium vaccinii*, *Malassezia globosa*, *Malassezia sympodialis*, *Pseudozyma antarctica*, *Pseudozyma hubeiensis*, *Sporisorium reilianum*, and *U. maydis*. The members of the species of other fungal groups analyzed were *Catenaria anguillulae* (Blastocladiomycota), *Batrachochytrium dendrobatidis* (Chytridiomycota), *Rozella allomyces* (Cryptomycota), *Conidiobolus coronatus* (Entomophthoromycotina), *Rhizophagus irregularis* and *Xylona heveae* (Glomeromycota), *Coemasia reversa* (Kickxellomycotina), *Backusella circina* and *Mucor circinelloides* (Mucoromycotina), *Orpinomyces* sp., and *Piromyces* sp. (Neocallimastigomycota).

The multiple sequence alignment carried out allowed us to conclude that the species representatives of Ascomycota and Basidiomycota, and at least one member of the rest of the fungal groups contain homologues of the following genes involved in polyamine metabolism: *ODC*, *SPDS*, *PAO*, and *SAMDC*. In addition, some Ascomycota and two Glomeromycota species contained the gene *SPMS*.

Regarding some motifs conserved in the different enzymes involved in polyamine biosynthesis, some results are relevant. Thus, sequences of amino acids conserved in Odcs determined by polymerase chain reaction (PCR) in some fungal species were reported by Torres-Guzman et al. (1996). These authors observed a conserved heptapeptide LDVGGGF in all fungal Odcs, clearly present in the sequences belonging to the phylum Ascomycota (Figure 4.2). Some substitutions of the valine residue were present in *Cladosporium fulvum*, *Blumeria graminis*, *Colletotrichum higginsianum*, and *Candida arabinofermentans* and in the four species representing the subphylum Taphrinomycotina. The same happened in the species representing the Basidiomycota phylum, where this region was conserved with substitution of the valine residue by isoleucine (Figure 4.3), and in the species representing the rest of the fungal groups (Figure 4.4). In general, it is interesting to notice some other conserved regions among the amino acid sequences, such as QH-RW-L, AVKCN, and GFDCAS.

The size of the Ascomycota Odc proteins ranges from 444 to 472 amino acids, with two exceptions, *Botrytis cinerea* with 531 amino acids and *Ascocoryne sarcoides* with 1534 amino acids (Figure 4.2). The *B. cinerea* protein sequence contains an *N*-terminal sequence extension of ≈90 amino acids with no obvious known function. The Odc homologous protein from *A. sarcoides* has homology at the *C*-terminal region with the rest of Odcs, while the sequence of ≈1000 amino acids present at the *N*-terminal region has homology with a putative mitochondrial presequence protease protein, a domain that is associated with the metallopeptidase family M16 (K.A. Johnson et al. 2006). Proteins of this kind are directed to the intermembrane space and the inner membrane of the mitochondria and are synthesized as larger proteins carrying *N*-terminal regions that are cleaved off (Gakh, Cavadini, and Isaya 2002). Whether this region in the Odc is involved in the determination of mitochondrial localization remains unknown.

In the case of the Odcs from Basidomycota, the proteins contained between 410 and 576 amino acids, with one extraordinary exception, *M. osmundae* with 1818 amino acids. In contrast to the previously mentioned examples, for the additional sequence ≈1300 amino acids present at de *N*-terminal region, no homology could be determined (Figure 4.3).

For the rest of species representing the fungal phylogeny, they essentially present the same features of the Odcs mentioned previously (Figure 4.4). However, the Odc from *R. allomyces* belonging to the Cryptomycota subphylum is a fused protein with homology at the *N*-terminal region with the *C*-terminal region of a 3-ketoacyl-Coenzyme A (CoA) thiolase. In fungi as well as in other organisms, thiolases are important to the peroxisomal fatty acid β-oxidation system, which has been reported as not essential for virulence in *Candida albicans* (Otzen et al. 2013), but its mutation in *Candida tropicalis* leads to an up-regulation of other peroxisomal proteins (Ueda et al. 2003). Whether this terminal region of the Odc from *R. allomyces* has

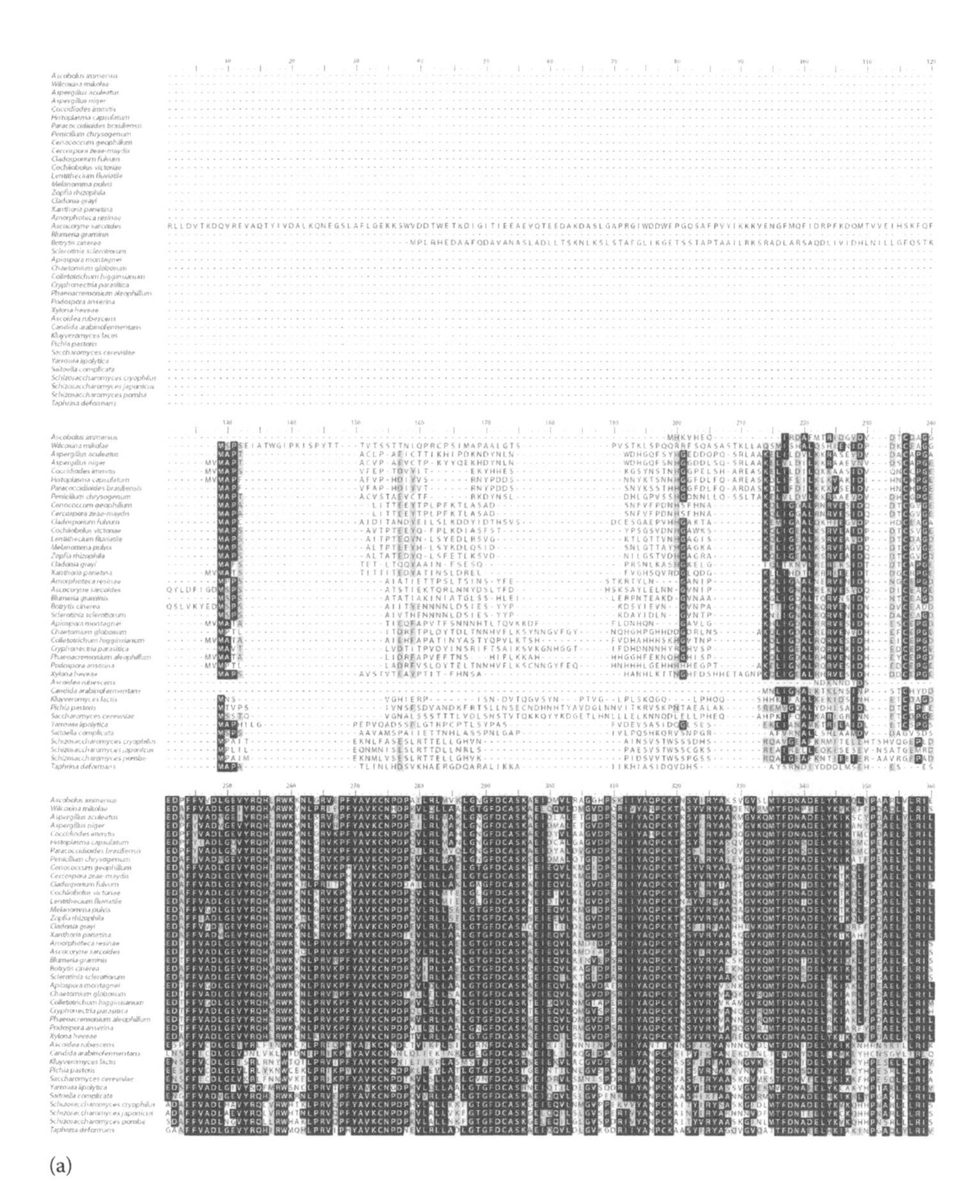

(a)

Figure 4.2 (a,b) Alignment of Odc protein sequences from fungi belonging to Ascomycota phylum. Amino acid residues that are identical in most of the sequences are highlighted in dark and similar residues in gray boxes. The amino acid sequences were obtained at the Genome Portal of the Department of Energy Joint Genome Institute (Nordberg, H., Cantor, M., Dusheyko, S., Hua, S., Poliakov, A., Shabalov, I., Smirnova, T., Grigoriev, I.V., Dubchak, I., *Nucleic Acids Res.*, 42, D26–D31, 2014) and alignment by ClustalW (Higgins, D.G., Thompson, J.D., Gibson, T.J., *Methods Enzymol.*, 266, 383–402, 1996). The fungal conserved heptapeptide LDVGGGF is marked by solid triangles above the sequence in panel b.

(Continued)

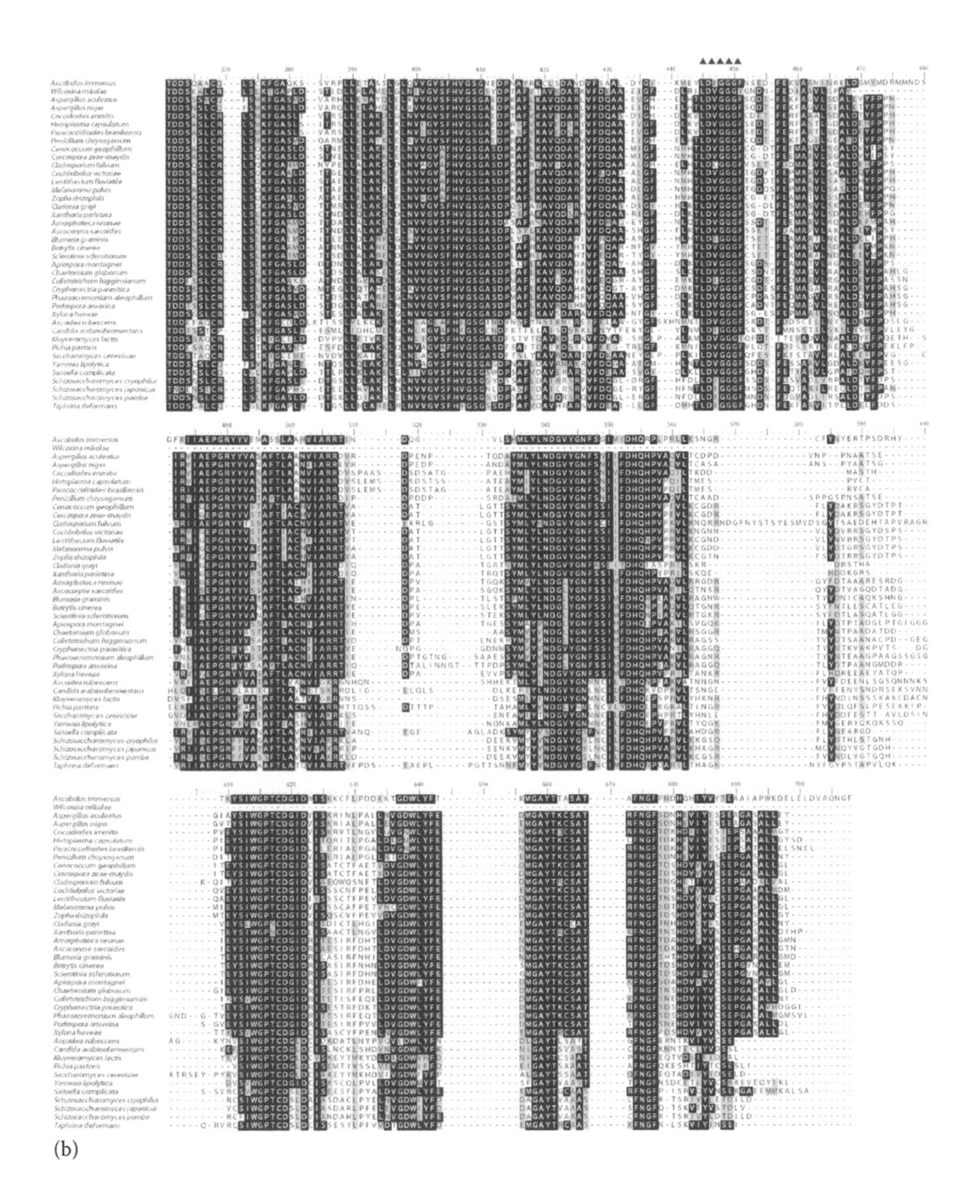

(b)

Figure 4.2 **(Continued)** (a,b) Alignment of Odc protein sequences from fungi belonging to Ascomycota phylum. Amino acid residues that are identical in most of the sequences are highlighted in dark and similar residues in gray boxes. The amino acid sequences were obtained at the Genome Portal of the Department of Energy Joint Genome Institute (Nordberg, H., Cantor, M., Dusheyko, S., Hua, S., Poliakov, A., Shabalov, I., Smirnova, T., Grigoriev, I.V., Dubchak, I., *Nucleic Acids Res.*, 42, D26–D31, 2014) and alignment by ClustalW (Higgins, D.G., Thompson, J.D., Gibson, T.J., *Methods Enzymol.*, 266, 383–402, 1996). The fungal conserved heptapeptide LDVGGGF is marked by solid triangles above the sequence in panel b.

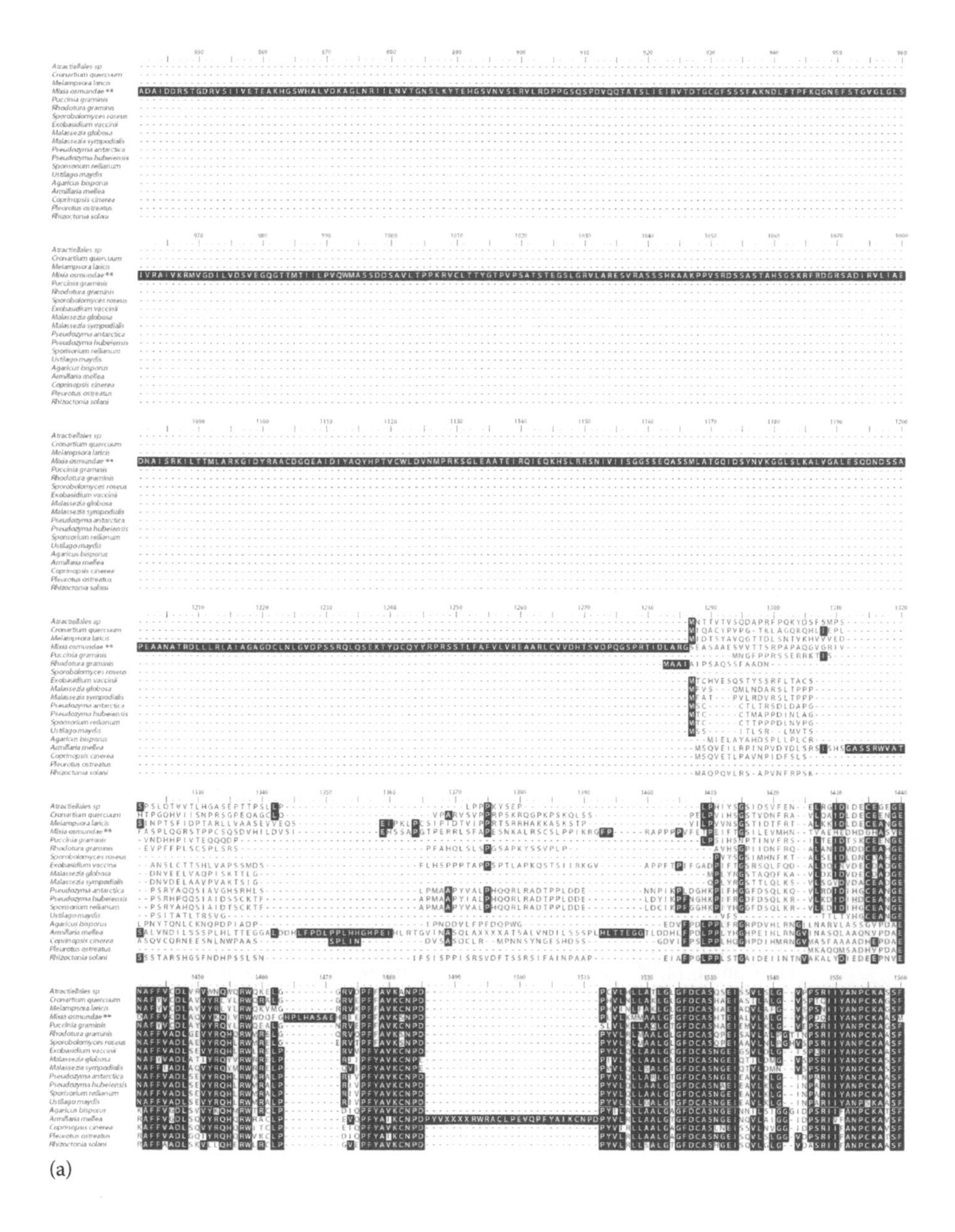

(a)

Figure 4.3 (a,b) Alignment of the Odc protein sequences of fungi belonging to the phylum Basidiomycota. The amino acid sequences were found at the Genome Portal of the Department of Energy Joint Genome Institute (Nordberg, H., Cantor, M., Dusheyko, S., Hua, S., Poliakov, A., Shabalov, I., Smirnova, T., Grigoriev, I.V., Dubchak, I., *Nucleic Acids Res.*, 42, D26–D31, 2014) and aligned by ClustalW (Higgins, D.G., Thompson, J.D., Gibson, T.J., *Methods Enzymol.*, 266, 383–402, 1996). The fungal conserved heptapeptide LDVGGGF is marked by solid triangles as in panel b. The *N*-terminal of *Mixia osmundae* corresponding to ≈1000 amino acids was removed.

(*Continued*)

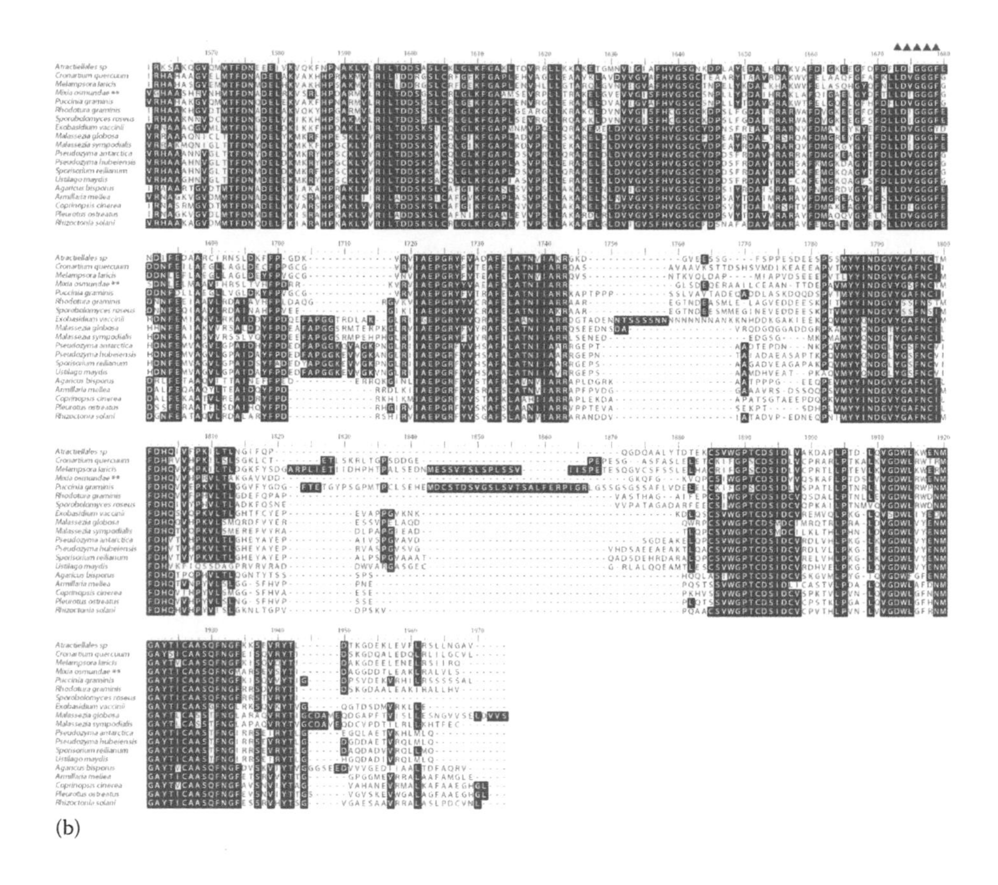

Figure 4.3 **(Continued)** (a,b) Alignment of the Odc protein sequences of fungi belonging to the phylum Basidiomycota. The amino acid sequences were found at the Genome Portal of the Department of Energy Joint Genome Institute (Nordberg, H., Cantor, M., Dusheyko, S., Hua, S., Poliakov, A., Shabalov, I., Smirnova, T., Grigoriev, I.V., Dubchak, I., *Nucleic Acids Res.*, 42, D26–D31, 2014) and aligned by ClustalW (Higgins, D.G., Thompson, J.D., Gibson, T.J., *Methods Enzymol.*, 266, 383–402, 1996). The fungal conserved heptapeptide LDVGGGF is marked by solid triangles as in panel b. The *N*-terminal of *Mixia osmundae* corresponding to ≈1000 amino acids was removed.

enzymatic activity remains unknown. It is thought that this kind of gene fusions to generate multidomain proteins is a mechanism of protein evolution, or that this phenomenon induces physical or structurally favorable configurations. Clustering of genes, and even their fusion, may occur between related or unrelated pathways (Yanai, Wolf, and Koonin 2002; Green et al. 2011). As examples, we may cite Odc fused with Samdc acting as a bifuntional protein reported in *Plasmodium falciparum* (Krause et al. 2000; Birkholtz et al. 2004), the existence in diverse bacterial phyla of gene fusion between Samdc and Spds that has been reported by Green et al. (2011), and the case of Spds from Basidiomycota cited further in Section 4.5.

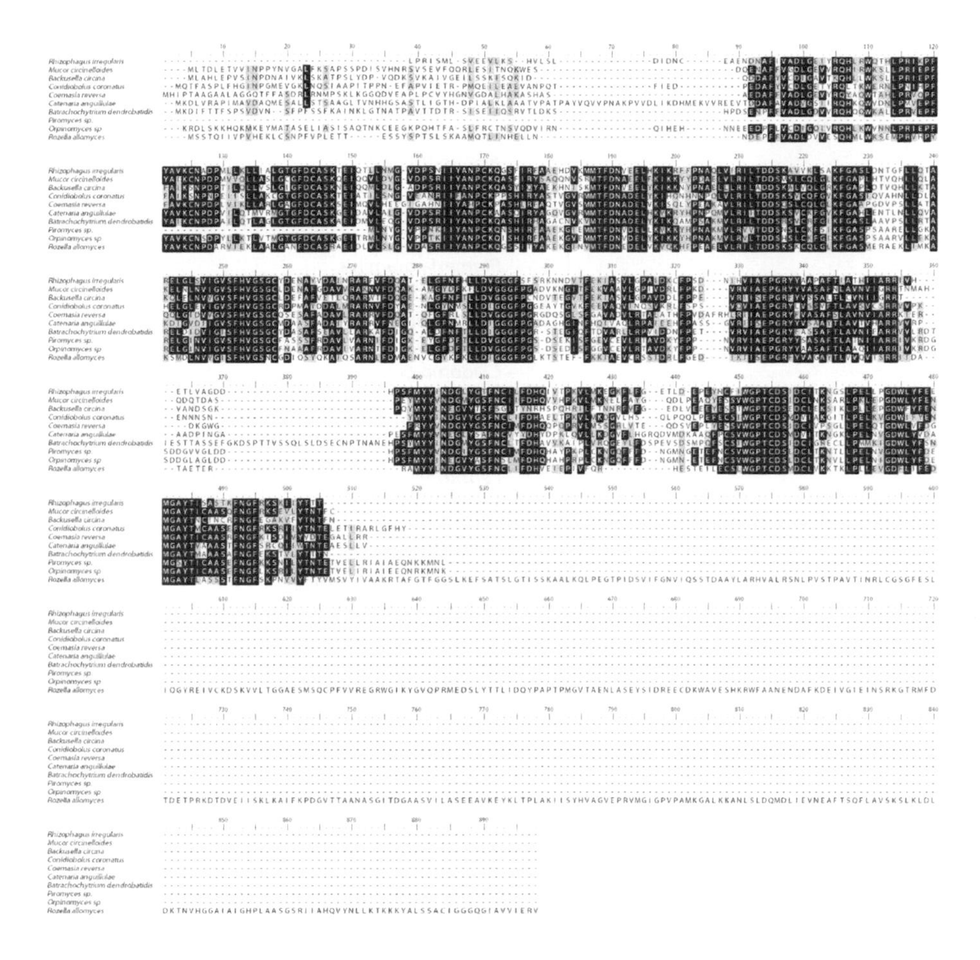

Figure 4.4 Alignment of the Odc protein sequences of fungi belonging to different phyla. The amino acid sequences were found at the Genome Portal of the Department of Energy Joint Genome Institute (Nordberg, H., Cantor, M., Dusheyko, S., Hua, S., Poliakov, A., Shabalov, I., Smirnova, T., Grigoriev, I.V., Dubchak, I., *Nucleic Acids Res.*, 42, D26–D31, 2014) and alignment by ClustalW (Higgins, D.G., Thompson, J.D., Gibson, T.J., *Methods Enzymol.*, 266, 383–402, 1996).

It is interesting to notice that the dendrogram generated from the alignments of Odc protein sequences showed a segregation of branches in agreement with the established phylogenetic relationships (Figures 4.5 through 4.7) (James et al. 2006; Hibbett et al. 2007).

Now, regarding Spds, in Ascomycota enzymes, the polypeptide contains between 247 and 342 amino acids (Haider et al. 2005). However, five of the total sequences analyzed here were shown to be larger, as observed in *A. fumigatus*, *A. niger*, *P. chrysogenum*, *A. sarcoides* (again, see previous discussion), and *S. sclerotiorum*, belonging to the *Eurotiomycetes* or *Leoteomycetes* classes, with 549, 542, 553, 588, and 614 amino acids, respectively (Figure 4.8). The sequence extension at

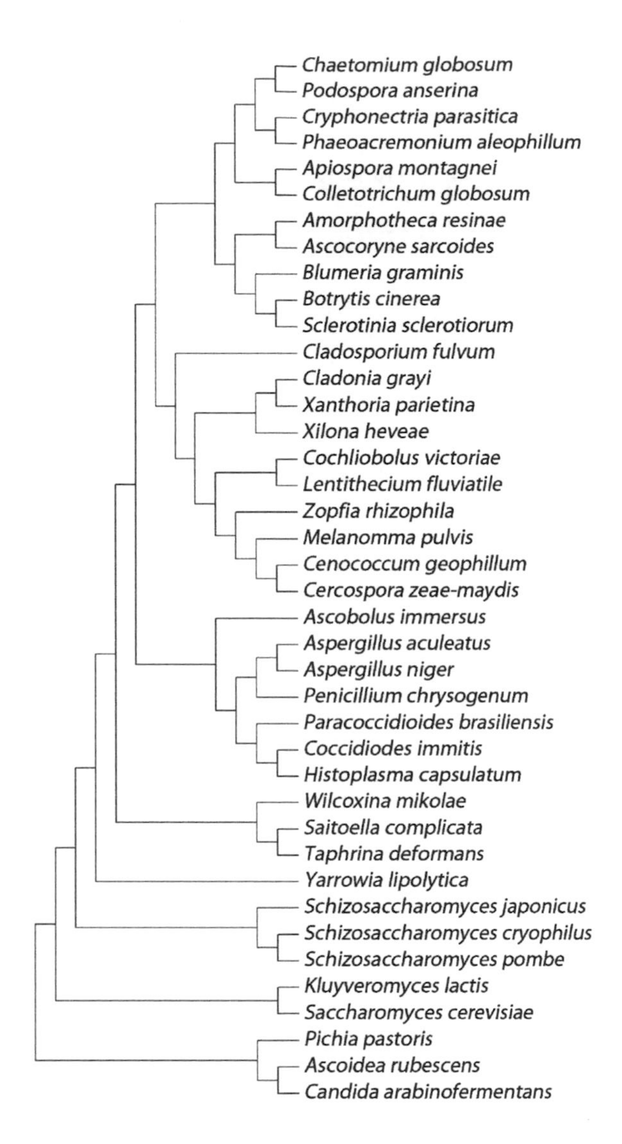

Figure 4.5 Phylogenetic tree showing the relationship of the deduced protein sequences of Odcs from Ascomycota. The amino acid sequences were found at the Genome Portal of the Department of Energy Joint Genome Institute (Nordberg, H., Cantor, M., Dusheyko, S., Hua, S., Poliakov, A., Shabalov, I., Smirnova, T., Grigoriev, I.V., Dubchak, I., *Nucleic Acids Res.*, 42, D26–D31, 2014), alignment by ClustalW (Higgins, D.G., Thompson, J.D., Gibson, T.J., *Methods Enzymol.*, 266, 383–402, 1996), and the tree was constructed by the neighbor-joining method with MEGA (ver. 3.1) program.

the *C*-terminal region has no known function. It is noticeable that the phylogenetic tree showed that both clusters group independently but had the same origin (Figure 4.9). Among the conserved motifs that could be identified, we may cite the putrescine aminopropyltransferases signature that corresponds to VIVIGGGDGGVLRE (Malone, Blumenthal, and Cheng 1995). Additionally, an *N*-terminal extension with

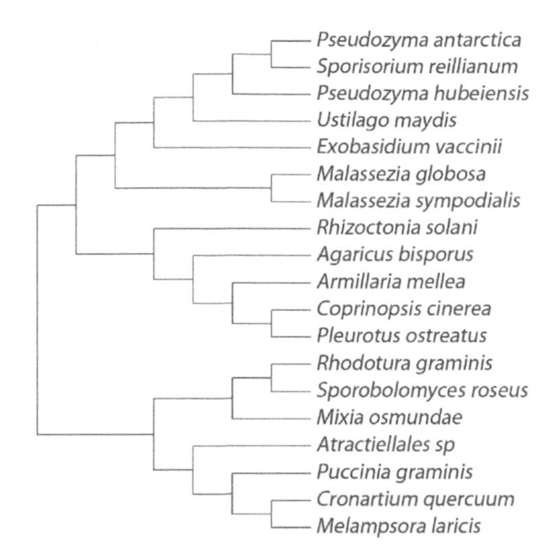

Figure 4.6 Phylogenetic tree showing the relationship of the deduced protein sequences of Odcs from Basidiomycota. The amino acid sequences were found at the Genome Portal of the Department of Energy Joint Genome Institute (Nordberg, H., Cantor, M., Dusheyko, S., Hua, S., Poliakov, A., Shabalov, I., Smirnova, T., Grigoriev, I.V., Dubchak, I., *Nucleic Acids Res.*, 42, D26–D31, 2014), alignment by ClustalW (Higgins, D.G., Thompson, J.D., Gibson, T.J., *Methods Enzymol.*, 266, 383–402, 1996), and the tree was constructed by the neighbor-joining method with MEGA (ver. 3.1) program.

unknown function is present in *C. immitis*, *P. brasiliensis*, *A. sarcoides*, and *S. sclerotiorum*. It is interesting to notice that some plants such as *Arabidopsis thaliana* contained this sequence, as well as *Plasmodium falciparum*, where the first 29 amino acid residues of Spds can be deleted and the expression of the enzyme is not affected (Haider et al. 2005; Dufe et al. 2007). Spds from Basidiomycota did not

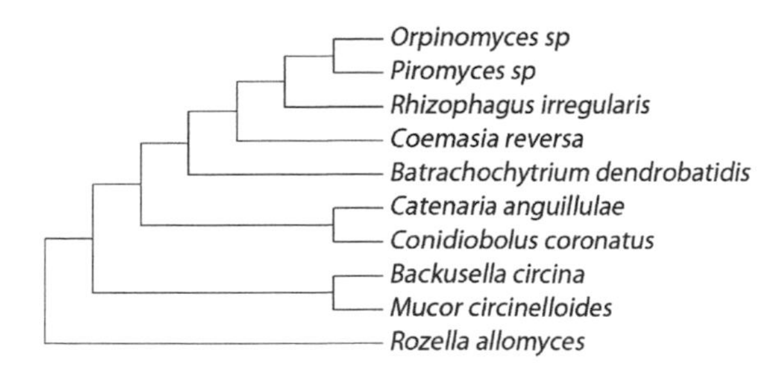

Figure 4.7 Phylogenetic tree analysis of Odcs from some representative fungi of Blasto cladiomycota, Chytridiomycota, Cryptomycota, Entomophthoromycota, Glomera mycota, Kickxellomycotina, Mucoromycotina, and Neocallimastigomycoya protein homologues. The amino acid sequences were found at the Genome Portal of the Department of Energy Joint Genome Institute (Nordberg, H., Cantor, M., Dusheyko, S., Hua, S., Poliakov, A., Shabalov, I., Smirnova, T., Grigoriev, I.V., Dubchak, I., *Nucleic Acids Res.*, 42, D26–D31, 2014), alignment by ClustalW (Higgins, D.G., Thompson, J.D., Gibson, T.J., *Methods Enzymol.*, 266, 383–402, 1996), and the tree constructed by the neighbor-joining method with MEGA (ver. 3.1) program.

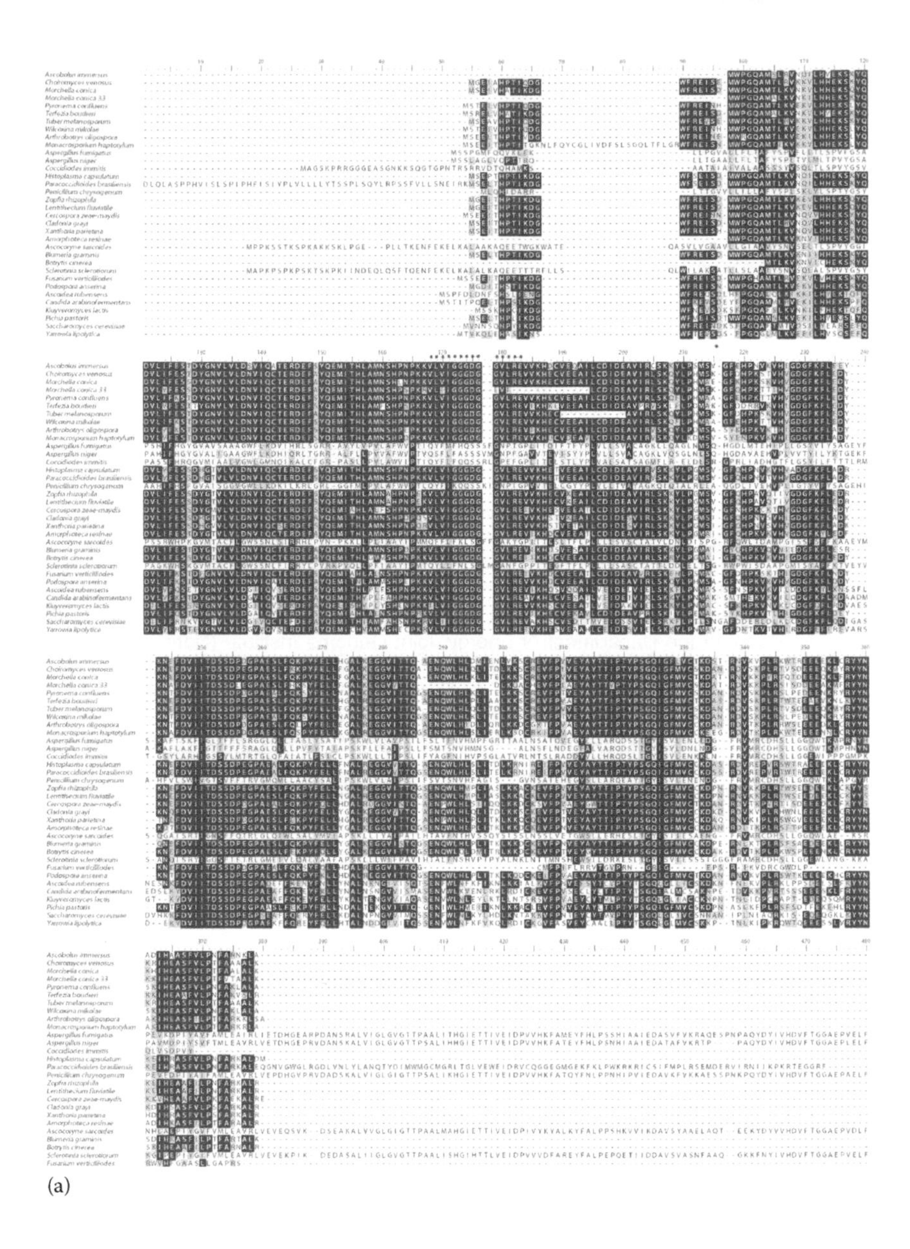

Figure 4.8 (a,b) Alignment of the Spds amino acid sequences of fungi belonging to phylum Ascomycota. The sequences have been aligned using the software ClustalW (Higgins, D.G., Thompson, J.D., Gibson, T.J., *Methods Enzymol.*, 266, 383–402, 1996). The sequences were extracted from the Genome Portal of the Department of Energy Joint Genome Institute (Nordberg, H., Cantor, M., Dusheyko, S., Hua, S., Poliakov, A., Shabalov, I., Smirnova, T., Grigoriev, I.V., Dubchak, I., *Nucleic Acids Res.*, 42, D26–D31, 2014). The sequences corresponding to putrescine aminopropyltransferase signature are marked with asterisks.

(*Continued*)

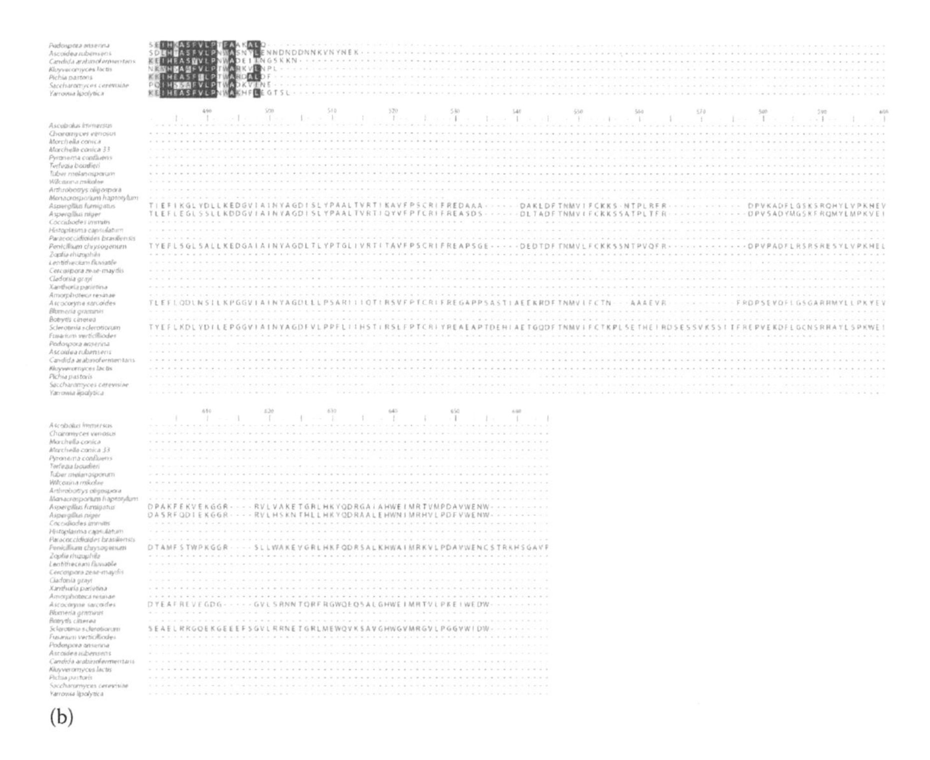

Figure 4.8 **(Continued)** (a,b) Alignment of the Spds amino acid sequences of fungi belonging to phylum Ascomycota. The sequences have been aligned using the software ClustalW (Higgins, D.G., Thompson, J.D., Gibson, T.J., *Methods Enzymol.*, 266, 383–402, 1996). The sequences were extracted from the Genome Portal of the Department of Energy Joint Genome Institute (Nordberg, H., Cantor, M., Dusheyko, S., Hua, S., Poliakov, A., Shabalov, I., Smirnova, T., Grigoriev, I.V., Dubchak, I., *Nucleic Acids Res.*, 42, D26–D31, 2014). The sequences corresponding to putrescine aminopropyltransferase signature are marked with asterisks.

present either any *N*-terminal sequence extension or *C*-terminal extensions. Rather, this is a special case because of its chimeric nature. In the Basidiomycota phylum, the *SPDS* gene is present as a fusion between two regions (Figures 4.10 and 4.11): at the *N*-terminal, there is a region homologous with Spds, and at the *C*-terminal, a region homologous with saccharopine dehydrogenase, an enzyme involved in lysine biosynthesis, separated by ≈18 amino acids. The second region has no an apparent methionine that could serve as a starting translational signal (Kingsbury et al. 2004; Leon-Ramirez et al. 2010). The rest of the representatives of the fungal kingdom contained an Spds protein sequence with no special feature; the shortest one was Spds from *R. irregularis*, with 273 amino acids, and the largest one from *Catenaria anguillulae*, with 329 amino acids. It must be noted that as described previously for other enzymes, the dendograms placed each one of enzymes belonging to the fungal organisms in accordance with their taxonomic relationships (Figures 4.12 and 4.13).

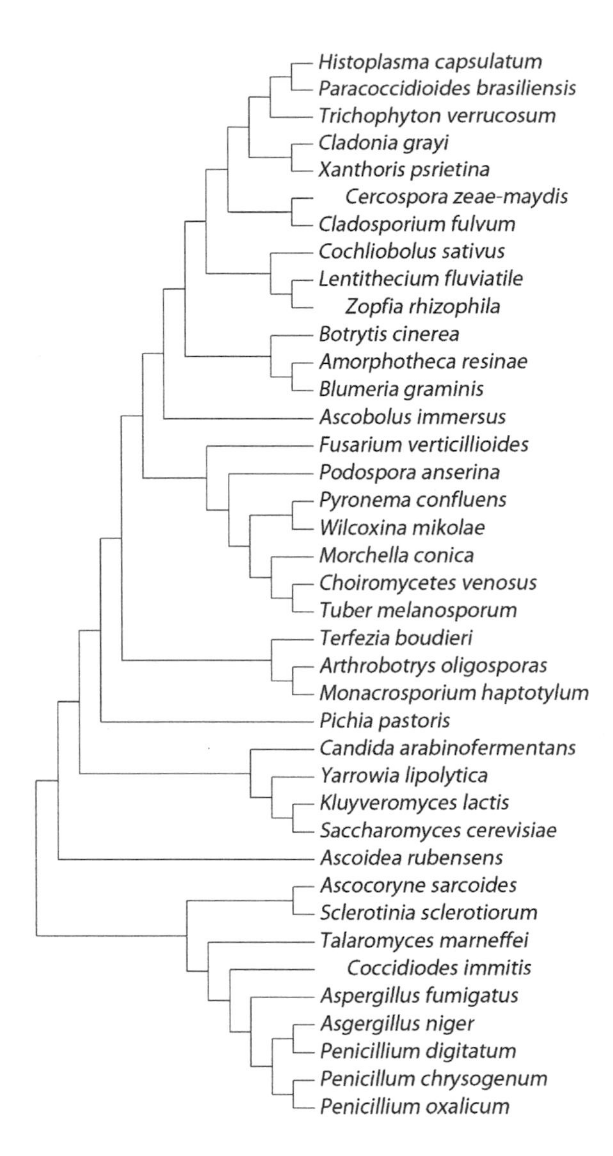

Figure 4.9 Phylogenetic tree showing the relationship of the deduced protein sequence of Spds from Ascomycota fungi. The amino acid sequences were found at the Genome Portal of the Department of Energy Joint Genome Institute (Nordberg, H., Cantor, M., Dusheyko, S., Hua, S., Poliakov, A., Shabalov, I., Smirnova, T., Grigoriev, I.V., Dubchak, I., *Nucleic Acids Res.*, 42, D26–D31, 2014), alignment by ClustalW (Higgins, D.G., Thompson, J.D., Gibson, T.J., *Methods Enzymol.*, 266, 383–402, 1996), and the tree was constructed by the neighbor-joining method with MEGA (ver. 3.1) program.

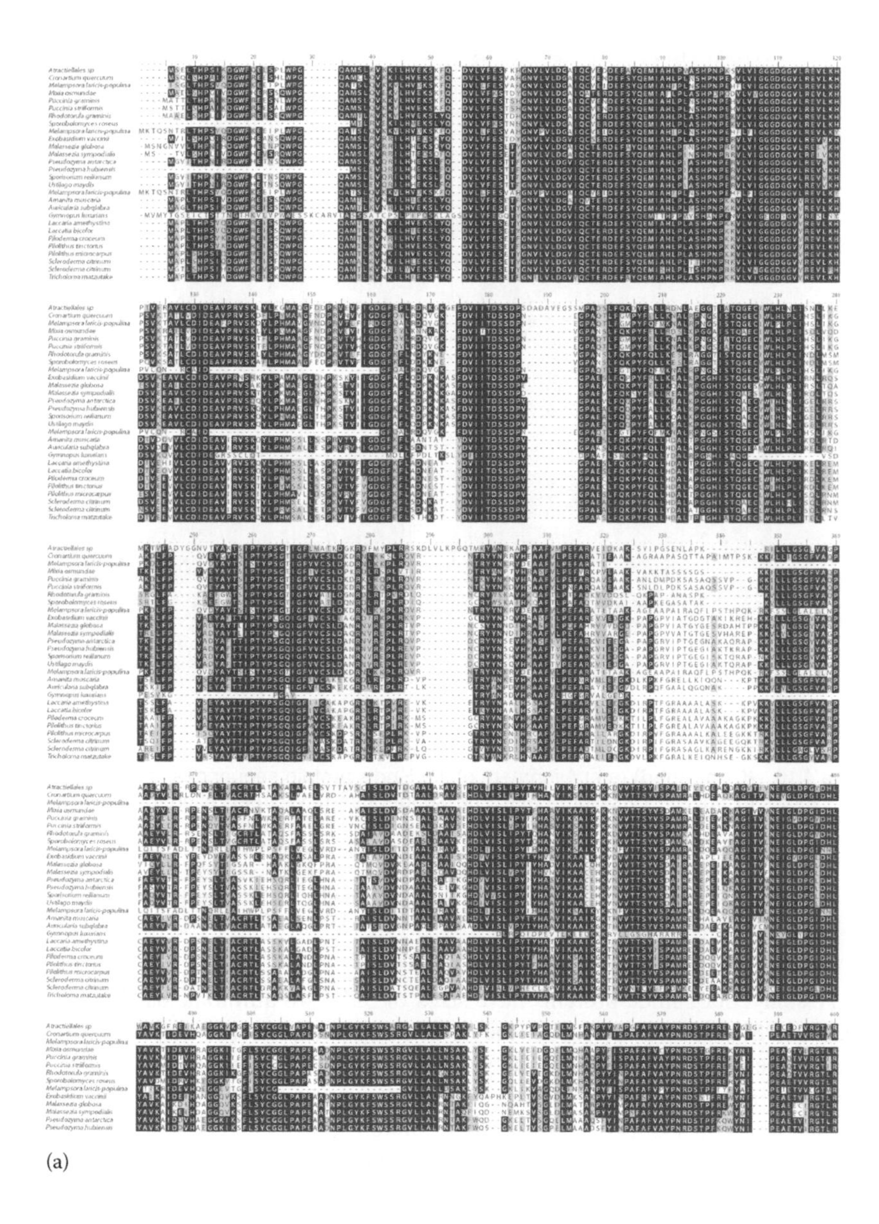

(a)

Figure 4.10 (a,b) Alignment of the Spd's protein sequences of fungi belonging to phylum Basidiomycota. The sequences have been aligned using the software ClustalW (Higgins, D.G., Thompson, J.D., Gibson, T.J., *Methods Enzymol.*, 266, 383–402, 1996). The sequences were extracted from the Genome Portal of the Department of Energy Joint Genome Institute (Nordberg, H., Cantor, M., Dusheyko, S., Hua, S., Poliakov, A., Shabalov, I., Smirnova, T., Grigoriev, I.V., Dubchak, I., *Nucleic Acids Res.*, 42, D26–D31, 2014).

(Continued)

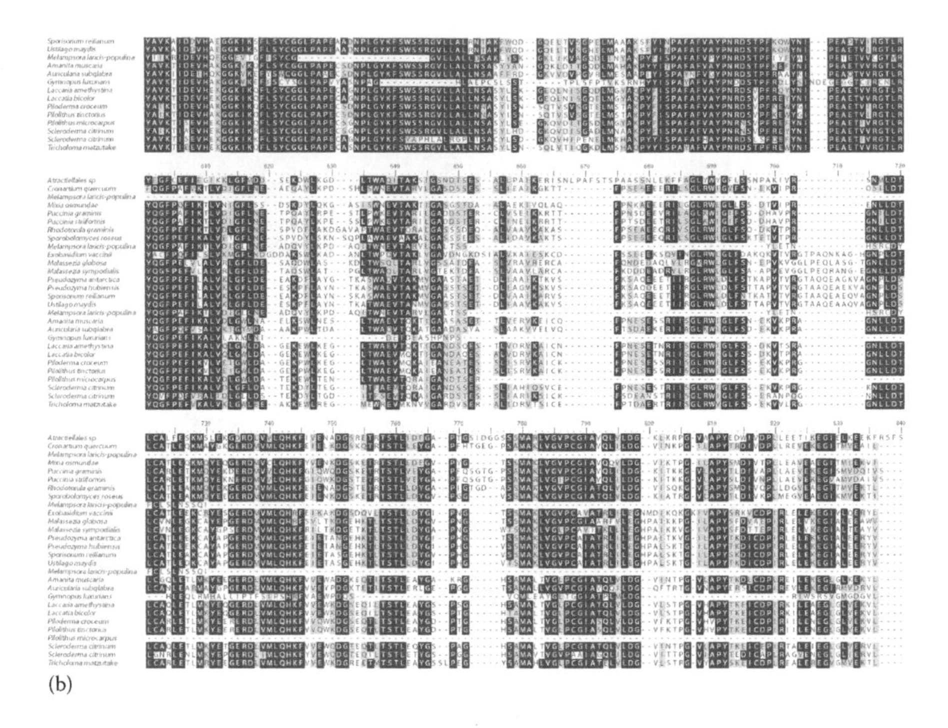

Figure 4.10 **(Continued)** (a,b) Alignment of the Spd's protein sequences of fungi belonging to phylum Basidiomycota. The sequences have been aligned using the software ClustalW (Higgins, D.G., Thompson, J.D., Gibson, T.J., *Methods Enzymol.*, 266, 383–402, 1996). The sequences were extracted from the Genome Portal of the Department of Energy Joint Genome Institute (Nordberg, H., Cantor, M., Dusheyko, S., Hua, S., Poliakov, A., Shabalov, I., Smirnova, T., Grigoriev, I.V., Dubchak, I., *Nucleic Acids Res.*, 42, D26–D31, 2014).

The Ascomycota Samdc proteins have sizes of 355–540 amino acids (Figure 4.14). Regarding Samdcs belonging to the Basidiomycota phylum, we found that the enzymes were larger, containing between 407 and 546 amino acids (Figure 4.15), but interestingly, the enzyme from *M. globosa* was much larger than any of them, with 622 amino acids. As mentioned in Chapter 3, the Samdc active enzymes are the result of an autocatalytic cleavage. It was observed that the conserved sequence YLLSESS occurs in all of the enzymes analyzed here. The underlined serine residue has been reported as the site of Samdc processing (Stanley and Pegg 1991), suggesting that the autocleavage reaction at this serine residue to generate the α and β subunits is conserved among the fungal kingdom. In the same line, putrescine seems to be necessary to enhance the processing and activity of Samdc (Pegg and Williams-Ashman 1969; Kameji and Pegg 1987), although it is not absolutely indispensable in all the systems, (see further discussion). Putrescine stimulation was completely abolished when a mutation was introduced in three acidic residues (Glu11, Glu178, and Glu256) of the human Samdc (Stanley and Pegg 1991). A comparison of fungal

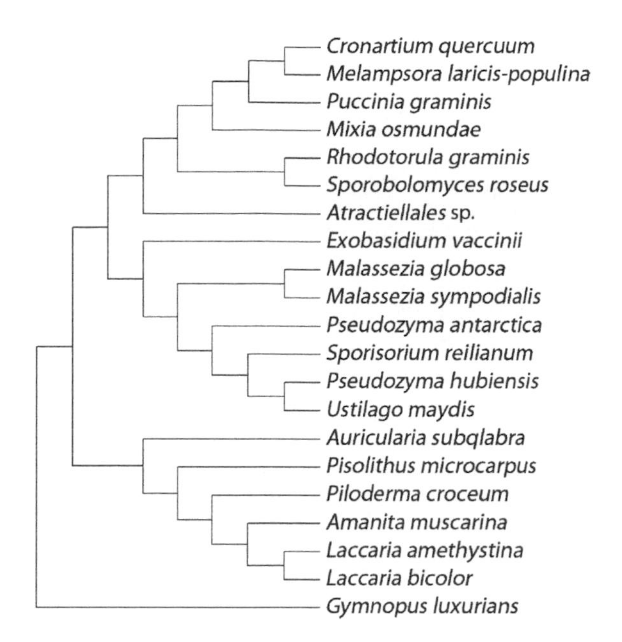

Figure 4.11 Phylogenetic tree showing the relationship of the deduced protein sequence of Spds proteins from species belonging to Basidiomycota. The amino acid sequences were found at the Genome Portal of the Department of Energy Joint Genome Institute (Nordberg, H., Cantor, M., Dusheyko, S., Hua, S., Poliakov, A., Shabalov, I., Smirnova, T., Grigoriev, I.V., Dubchak, I., *Nucleic Acids Res.*, 42, D26–D31, 2014), alignment by ClustalW (Higgins, D.G., Thompson, J.D., Gibson, T.J., *Methods Enzymol.*, 266, 383–402, 1996), and the was tree constructed by the neighbor-joining method with MEGA (ver 3.1) program.

Samdc sequences showed that the regions containing the glutamic acid residues and the aspartic acid needed for stimulation and processing of the Samdc by putrescine are conserved: PFEGPEKLLEI, DETLEIIMTGL, and TLFVTPE–SYASFE (Figures 4.14 and 4.15). However, in *N. crassa*, putrescine had no effect on the stimulation of processing of the proenzyme, despite the conservation of the same residues; putrescine was involved only in stimulation of the activity of Samdc (Hoyt et al. 2000). In addition, the presence of the three acidic residues in plant Samdc enzymes has been reported in plants, but none of them appear to require putrescine for their activation (Xiong et al. 1997).

The multiple alignment of the rest of fungal Samdc sequences revealed that they contain 310 to 581 amino acids, although *Batrachochytrium dendrobatidis* Samdc was much longer than any of them, with 771 amino acids, and with no homology with any Samdc proenzyme at the *C*-terminal region (Figure 4.16). An analysis of the *C*-terminal region indicated the presence of an F-box domain, a sequence known as a site of protein–protein interaction (Kipreos and Pagano 2000).

Putative fungal homologs of Spms were identified in some representatives of the following Ascomycota groups: Dothideomycetes, Eurotiomycetes, Leotiomycetes,

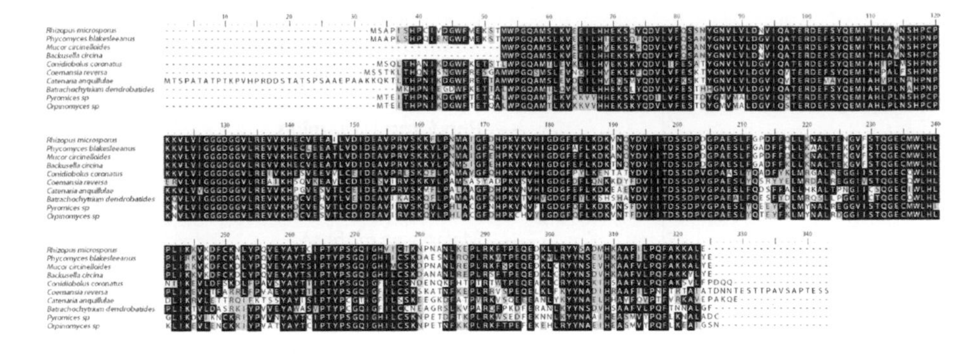

Figure 4.12 Alignment of Spds protein sequences of fungi belonging to different phyla. The sequences have been aligned using the software ClustalW (Higgins, D.G., Thompson, J.D., Gibson, T.J., *Methods Enzymol.*, 266, 383–402, 1996). The sequences were extracted from the Genome Portal of the Department of Energy Joint Genome Institute (Nordberg, H., Cantor, M., Dusheyko, S., Hua, S., Poliakov, A., Shabalov, I., Smirnova, T., Grigoriev, I.V., Dubchak, I., *Nucleic Acids Res.*, 42, D26–D31, 2014).

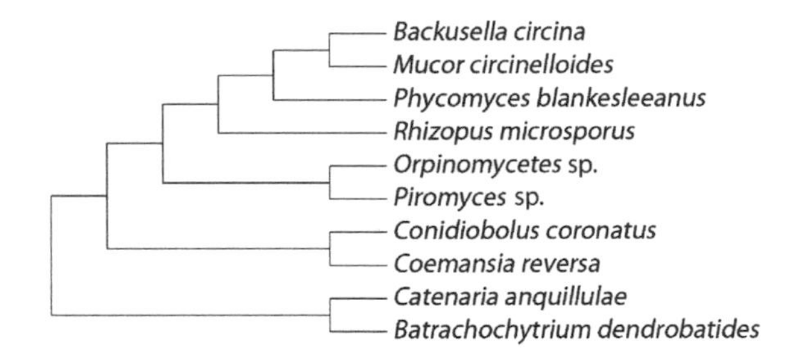

Figure 4.13 Phylogenetic tree of Spds protein from organisms belonging to Mucoromycotina, Entomophthoramycotina, Kickxellomycotina, Blastocladiomycota, Chytridiomycota, and Neocallimastigomycota. The amino acid sequences were found at the Genome Portal of the Department of Energy Joint Genome Institute (Nordberg, H., Cantor, M., Dusheyko, S., Hua, S., Poliakov, A., Shabalov, I., Smirnova, T., Grigoriev, I.V., Dubchak, I., *Nucleic Acids Res.*, 42, D26–D31, 2014), alignment by ClustalW (Higgins, D.G., Thompson, J.D., Gibson, T.J., *Methods Enzymol.*, 266, 383–402, 1996), and the tree was constructed by the neighbor-joining method with MEGA (ver 3.1) program.

Pezizomycetes, Saccharomycotina, Sordariomycetes, Xilomicetes, and, interestingly, in at least one species of Glomeromycota, a phylum established recently that was removed from Zygomycota and is thought to be a divergent group sharing an ancestor with Ascomycota and Basidiomycota (Schubler, Schwarzott, and Walker 2001) (Table 4.1, Figure 4.17). Since Basidiomycota fungi lack Spms, it might be reasonable to conclude a closer relationship of Glomeromycota with Ascomycota phylum. Nevertheless, since only one Glomeromycota species showed to possibly contain Spms, it would be wise to analyze whether this is not an isolated case and try to find additional elements that support the previously mentioned hypothesis.

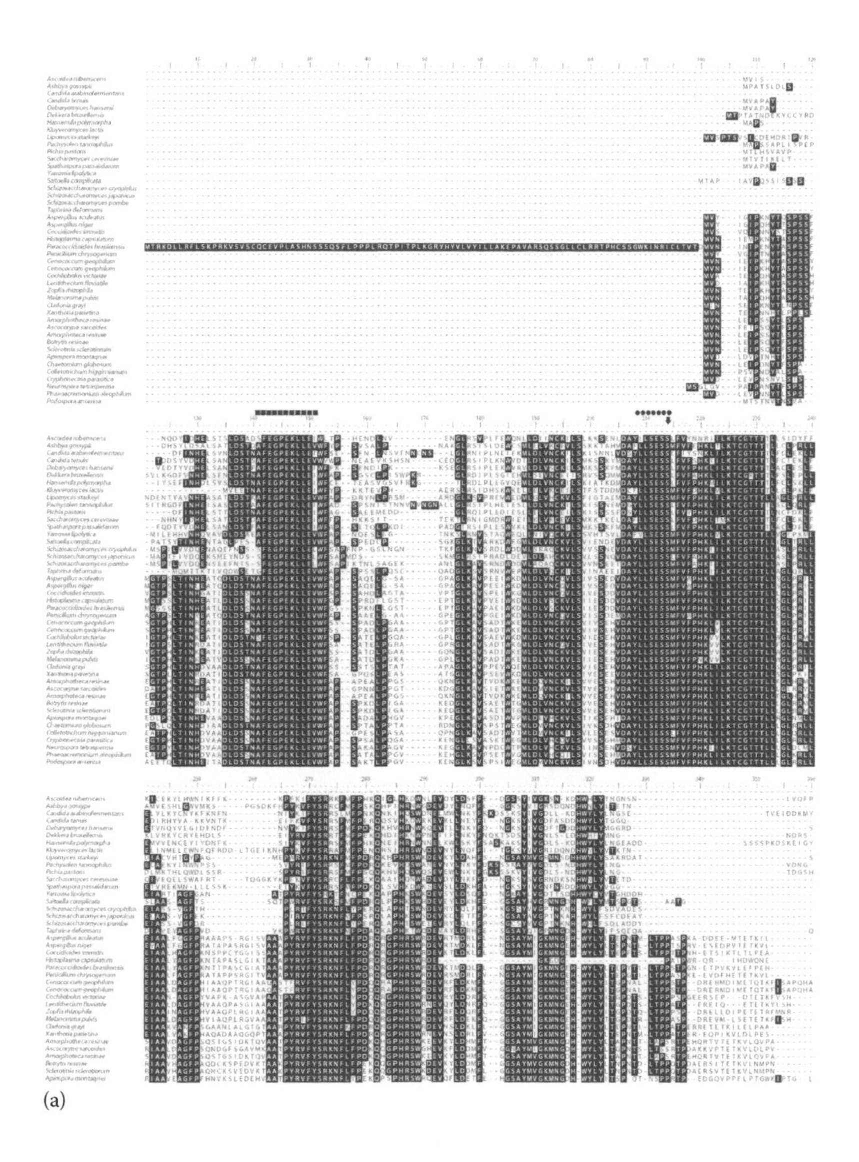

(a)

Figure 4.14 (a–c) Alignment of the Samdc proenzyme sequences of fungi belonging to phylum Ascomycota. The sequences have been aligned using the software ClustalW (Higgins, D.G., Thompson, J.D., Gibson, T.J., *Methods Enzymol.*, 266, 383–402, 1996). The sequences were extracted from the Genome Portal of the Department of Energy Joint Genome Institute (Nordberg, H., Cantor, M., Dusheyko, S., Hua, S., Poliakov, A., Shabalov, I., Smirnova, T., Grigoriev, I.V., Dubchak, I., *Nucleic Acids Res.*, 42, D26–D31, 2014). The sequence YLLSESS is highlighted with solid circles, and into this sequence, the serine under the solid arrow corresponds to the residue that suffers the autocleavage reaction to generate the α and β subunits. Regions involved in stimulation and processing of the Samdc by putrescine (PFEGPEKLLEI, DETLEIIMTGL, and TLFVTPE–SYASFE) are marked with solid squares above sequences.

(*Continued*)

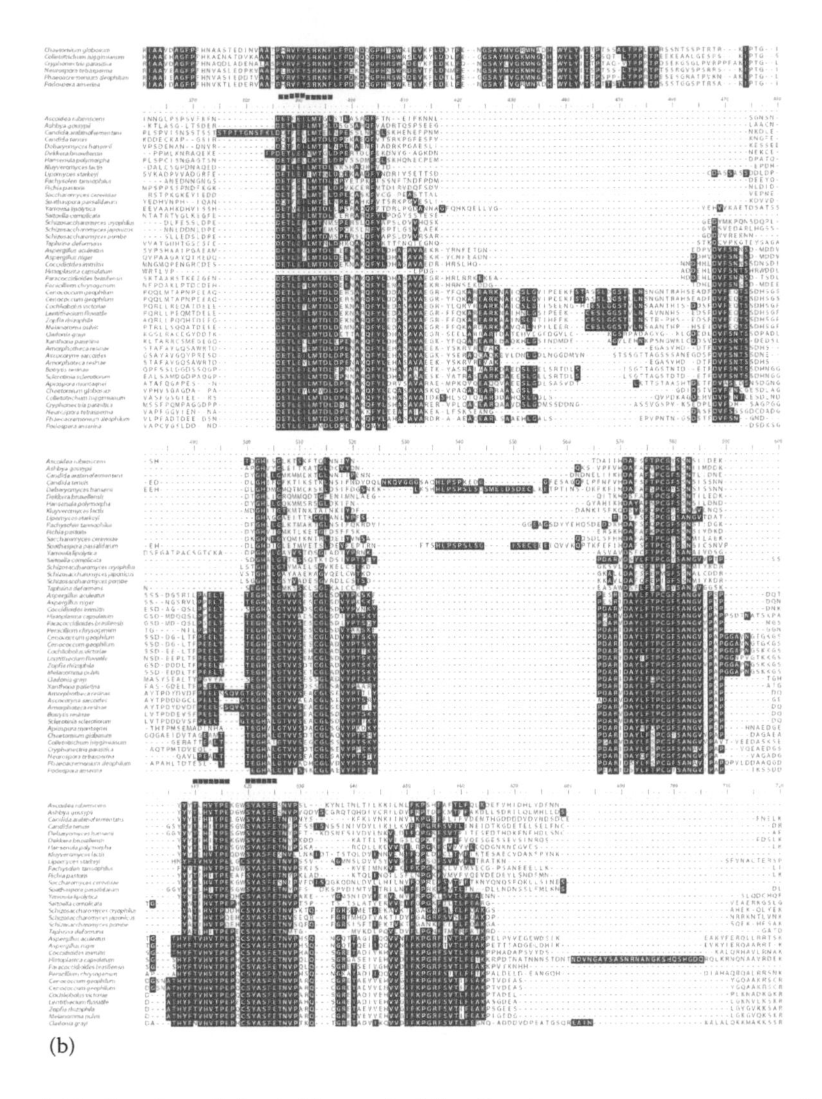

(b)

Figure 4.14 (Continued) (a–c) Alignment of the Samdc proenzyme sequences of fungi belonging to phylum Ascomycota. The sequences have been aligned using the software ClustalW (Higgins, D.G., Thompson, J.D., Gibson, T.J., *Methods Enzymol.*, 266, 383–402, 1996). The sequences were extracted from the Genome Portal of the Department of Energy Joint Genome Institute (Nordberg, H., Cantor, M., Dusheyko, S., Hua, S., Poliakov, A., Shabalov, I., Smirnova, T., Grigoriev, I.V., Dubchak, I., *Nucleic Acids Res.*, 42, D26–D31, 2014). The sequence YLLSESS is highlighted with solid circles, and into this sequence, the serine under the solid arrow corresponds to the residue that suffers the autocleavage reaction to generate the α and β subunits. Regions involved in stimulation and processing of the Samdc by putrescine (PFEGPEKLLEI, DETLEIIMTGL, and TLFVTPE–SYASFE) are marked with solid squares above sequences.

(Continued)

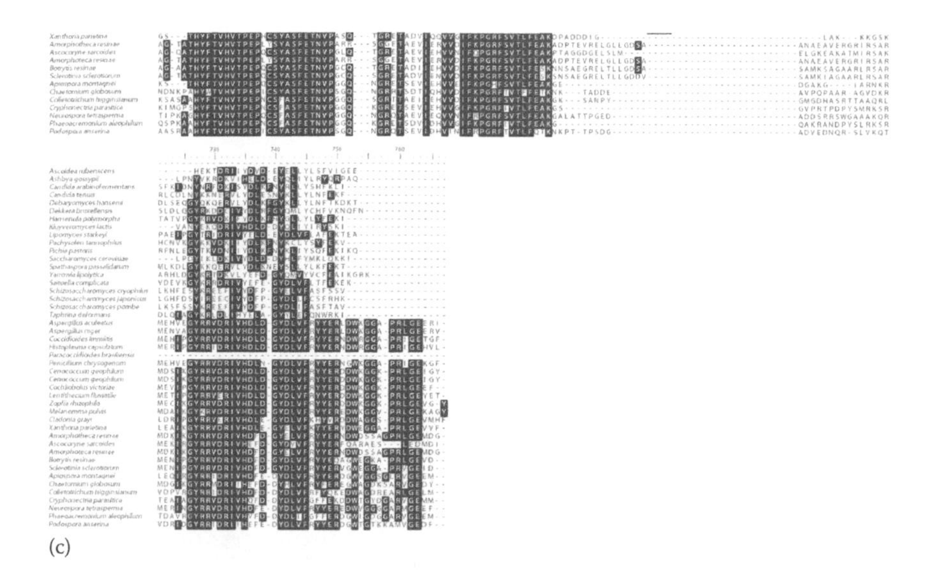

(c)

Figure 4.14 (Continued) (a–c) Alignment of the Samdc proenzyme sequences of fungi belonging to phylum Ascomycota. The sequences have been aligned using the software ClustalW (Higgins, D.G., Thompson, J.D., Gibson, T.J., *Methods Enzymol.*, 266, 383–402, 1996). The sequences were extracted from the Genome Portal of the Department of Energy Joint Genome Institute (Nordberg, H., Cantor, M., Dusheyko, S., Hua, S., Poliakov, A., Shabalov, I., Smirnova, T., Grigoriev, I.V., Dubchak, I., *Nucleic Acids Res.*, 42, D26–D31, 2014). The sequence YLLSESS is highlighted with solid circles, and into this sequence, the serine under the solid arrow corresponds to the residue that suffers the autocleavage reaction to generate the α and β subunits. Regions involved in stimulation and processing of the Samdc by putrescine (PFEGPEKLLEI, DETLEIIMTGL, and TLFVTPE–SYASFE) are marked with solid squares above sequences.

It has been suggested that spermine emerged during evolution together with the cell nucleus, but it is not clear what its real importance is, although an evolutionary advantage has been strongly suggested because of its conservation in so many organisms (Seiler 2004; Minguet et al. 2008). Nevertheless, as was indicated in Chapter 3, it was demonstrated that spermine-lacking mutants of *S. cerevisiae* did not present an obvious phenotype (Hamasaki-Katagiri et al. 1998). The suggestion that spermine might be involved in the formation of polyamine complex regulatory mechanism, as well as in some interconversion reaction, would nevertheless favor the hypothesis of their importance in some organisms.

Regarding enzymes involved in back-conversion of polyamines, it is an interesting observation that they possess low levels of sequence protein homology among themselves (Angus-Hill et al. 1999); accordingly, the construction of a phylogenetic analysis was not feasible. In this aspect, it has been reported that Ssats form a group member of the *N*-acetyltransferase family, which includes enzymes such as Gcn5, a

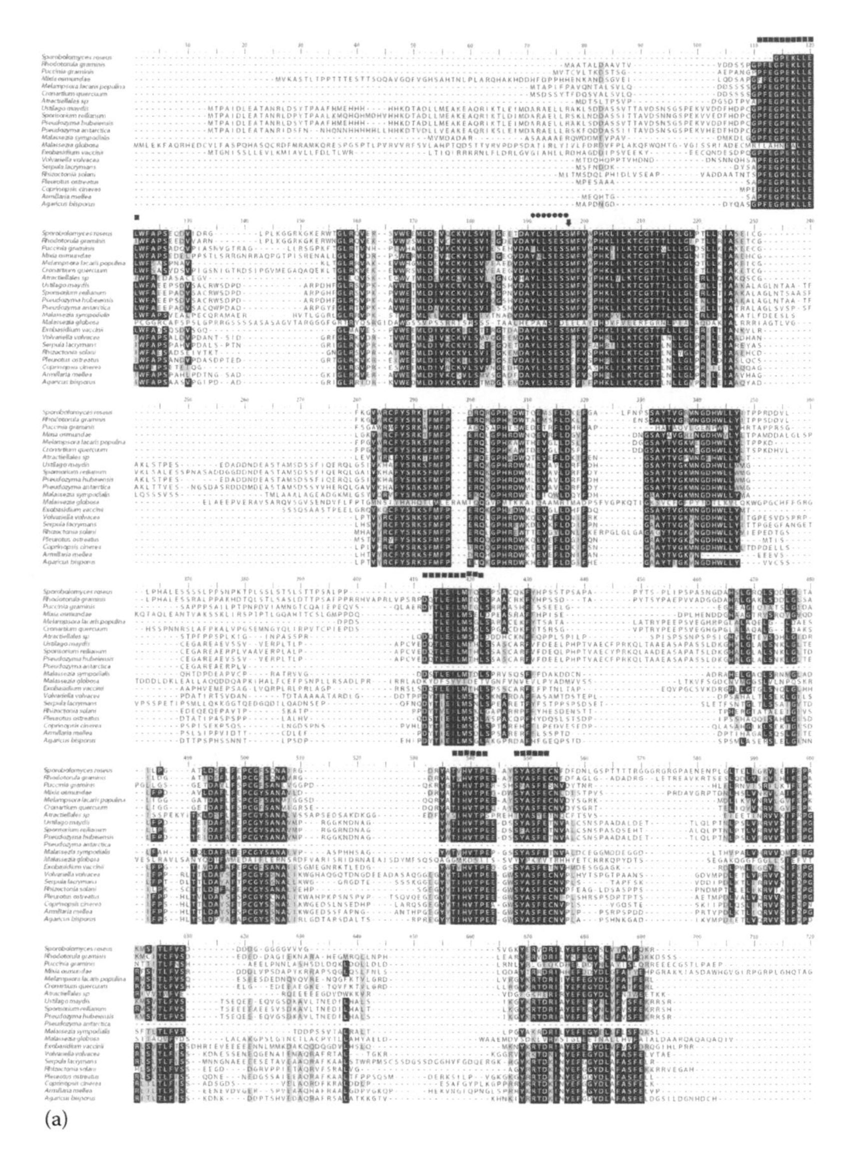

Figure 4.15 (a,b) Alignment of the Samdc proenzyme sequences of fungi belonging to phylum Basidiomycota. The amino acid sequences were found at the Genome Portal of the Department of Energy Joint Genome Institute (Nordberg, H., Cantor, M., Dusheyko, S., Hua, S., Poliakov, A., Shabalov, I., Smirnova, T., Grigoriev, I.V., Dubchak, I., *Nucleic Acids Res.*, 42, D26–D31, 2014). The sequence YLLSESS is highlighted with solid circles, and into this sequence, the serine under the solid arrow corresponds to the residue that suffers the autocleavage reaction to generate α and β subunits. Regions involved in stimulation and processing of the Samdc by putrescine (PFEGPEKLLEI, DETLEIIMTGL, TLFVTPE–SYASFE) are marked with solid squares above sequences.

(*Continued*)

(b)

Figure 4.15 **(Continued)** (a,b) Alignment of the Samdc proenzyme sequences of fungi belonging to phylum Basidiomycota. The amino acid sequences were found at the Genome Portal of the Department of Energy Joint Genome Institute (Nordberg, H., Cantor, M., Dusheyko, S., Hua, S., Poliakov, A., Shabalov, I., Smirnova, T., Grigoriev, I.V., Dubchak, I., *Nucleic Acids Res.*, 42, D26–D31, 2014). The sequence YLLSESS is highlighted with solid circles, and into this sequence, the serine under the solid arrow corresponds to the residue that suffers the autocleavage reaction to generate α and β subunits. Regions involved in stimulation and processing of the Samdc by putrescine (PFEGPEKLLEI, DETLEIIMTGL, TLFVTPE–SYASFE) are marked with solid squares above sequences.

histone acetyltransferase. In fact, some of the members of this family present high identity at the protein level with Ssat, although logically, they could not be involved in polyamine metabolism (Coleman et al. 2004).

As already indicated, Paos play an important role in the back-conversion of polyamines and the catabolism of polyamines. These enzymes are widely distributed in fungi, plants, and bacteria, as well as in mammalian cells. Pao sequences present at the *N*-terminal region the flavin-adenine dinucleotide-binding motif G-A-G-I-A-G (Dailey and Dailey 1998; Tavladoraki et al. 1998; Nishikawa et al. 2000) and the amino acid sequence –S-K-L at the *C*-terminal region, corresponding to the peroxisomal targeting signal sequence (Wallace 1998; reviewed by Sebela et al. 2001). These are two of the most important conserved features, while the rest of the protein sequence does not show significant homology between them.

In relation to the uncommon polyamines, additional studies will be necessary to determine how widespread they are in the fungal kingdom and to analyze their physiological roles and the mechanisms involved in their synthesis. Interestingly, in bacteria, the presence of sym-norspermidine has been considered a common characteristic of the genus *Vibrio* (Yamamoto et al. 1991).

4.4 POSSIBLE EVOLUTION OF THE POLYAMINE BIOSYNTHETIC PATHWAYS IN FUNGI

It is known that with some variants, specifically in the synthesis of putrescine, the orthodox polyamine metabolic pathway is present in all organisms. Regarding fungi, the presence of Odc, Spds, Samdc, and Pao enzymes in all fungal genomes analyzed

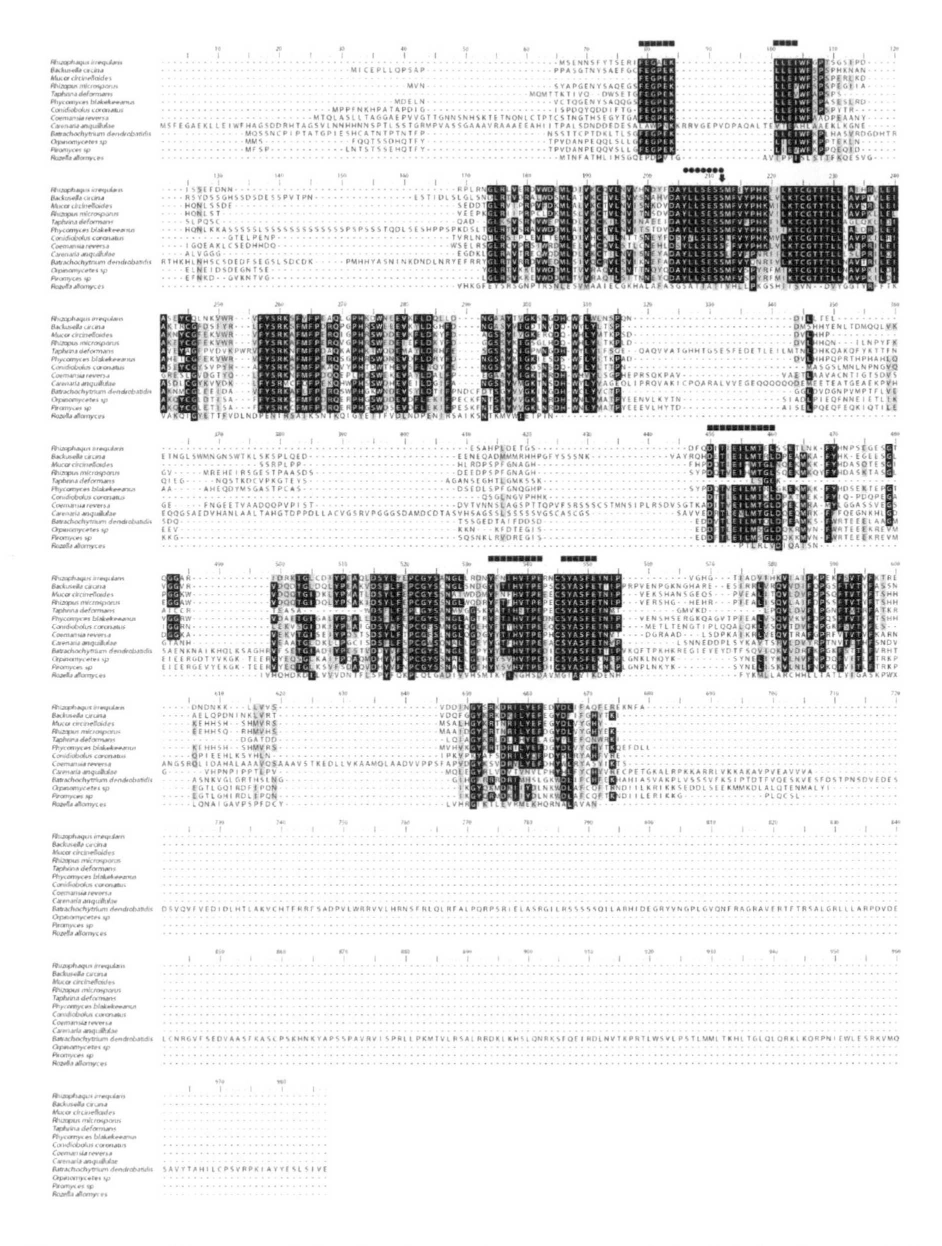

Figure 4.16 Alignment of the Samdc protein sequences of fungi belonging to phyla Glomeromycota, Blastocladiomycota, Chytridiomycota, Neocallimastigomycota, and Cryptomycota, as well as subphylum Mucoromycotina, Entomophthoromycotina, and Kickxellomycotina. The sequences have been aligned using the software ClustalW (Higgins, D.G., Thompson, J.D., Gibson, T.J., *Methods Enzymol.*, 266, 383–402, 1996). The sequences were extracted from the Genome Portal of the Department of Energy Joint Genome Institute (Nordberg, H., Cantor, M., Dusheyko, S., Hua, S., Poliakov, A., Shabalov, I., Smirnova, T., Grigoriev, I.V., Dubchak, I., *Nucleic Acids Res.*, 42, D26–D31, 2014).

Table 4.1 Fungal Species Where Putative Spermine Synthases Were Identified

Fungal Taxa	Organism	Protein ID[a]
Dothideomycetes	*Cercospora zeae maydis*	82864
	Cochliobolus sativus	31026
Eurotiomycetes	*Aspergillus aculeatus*	49136
	Aspergillus niger	1148152
	Coccidioides immitis	3120
	Penicillium chrysogenum	34296
	Rhizophagus irregularis	346928
Leotiomycetes	*Blumeria graminis*	19742
	Sclerotinia sclerotiorum	3826
Pezizomycetes	*Choiromycetes venosus*	1841192
	Morchella conica	509448
Xylomicetes	*Pyronema confluens*	11977
	Xylona heveae	246406
Saccharomycotina	*Pichia pastoris*	40023
	Saccharomyces cerevisiae	4167
	Yarrowia lipolytica	69510
Sordariomycetes	*Apiospora montagnei*	619293
	Chaetomium globosum	16951
	Colletotrichum higginsianum	8774
	Cryphonectria parasitica	103380
	Magnaporthe grisea	119385
	Phaeoacremonium aleophilum	3541
Glomeromycota	*Rhizophagus irregularis*	228408

Note: Unpublished data from Laura Valdés-Santiago.
[a] Data were obtained by BLASTP at http://genome.jgi.doe.gov/programs/fungi/index.jsf.

thus far (see previous discussion) indicates that the genes encoding these enzymes appeared early in the evolution of the kingdom. However, phylogenetical interconnection of the aminotransferases (Spds, Spms) and their specificity, as well as the absence of thermospermine synthase, has suggested that they might have originated from an event of diversification of a preexisting function (Minguet et al. 2008). Fungi do not contain thermospermine, as occurs in some bacteria and plants, a characteristic that may be explained by the loss of the gene encoding thermospermine synthase or that plants obtained it by horizontal gene transfer from bacteria (Minguet et al. 2008). The case of Spms is noticeable; phylogenetic studies reveal that it emerged independently in animals, the Ascomycota and, possibly, Glomeromycota groups in fungi, and plants. Minguet et al. (2008) suggested that the gene encoding Spms evolved from the gene encoding Spds by duplication and change of functions in each phylum. In the context of fungi, the presence of Spms is interesting since its presence is confined to only some taxa (Table 4.1; Pegg and Michael 2010). In this point, several questions arise. Why is Spms not present in some fungal taxa such as Basidiomycota and Zygomycota? Was it lost during evolution, or was never acquired? These questions have not been solved yet (see further discussion). Interestingly, only in vertebrates was

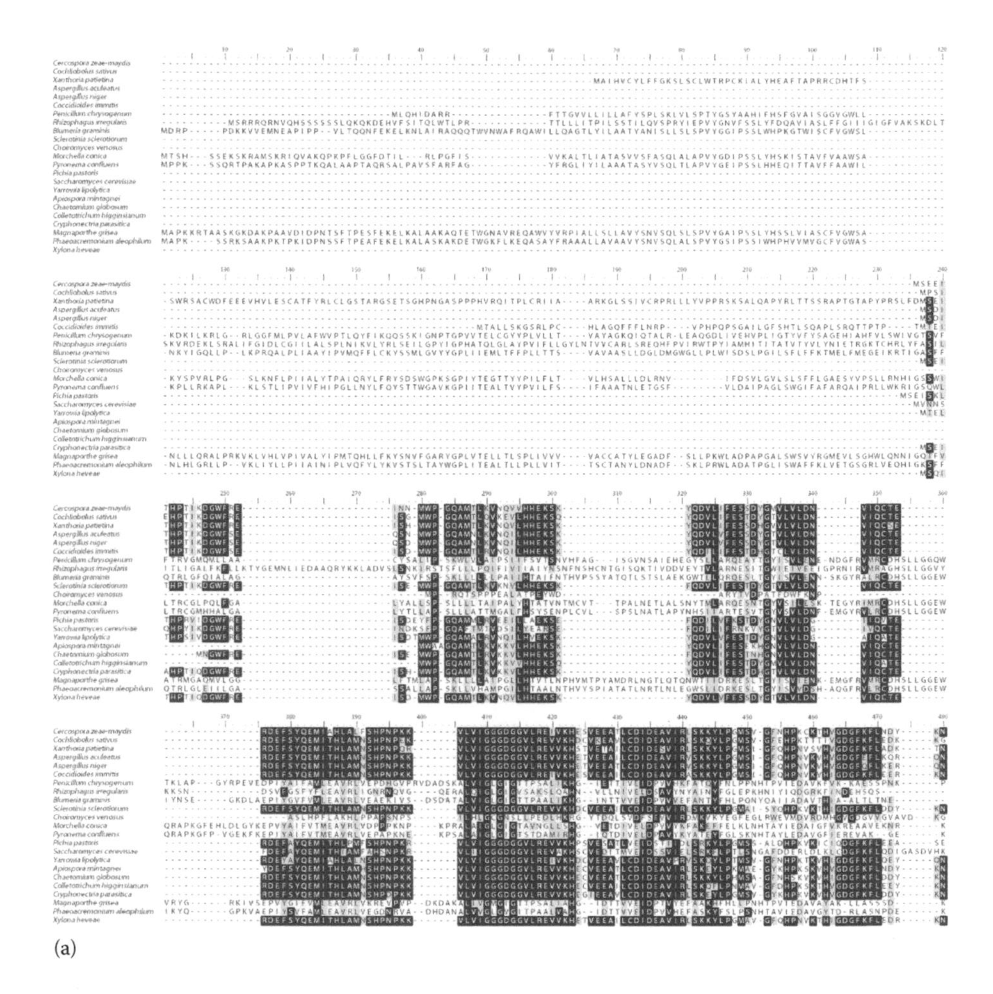

(a)

Figure 4.17 (a,b) Alignment of the Spm protein sequences of some fungal member belonging to Dothideomycetes, Eurotiomycetes, Glomeramycota, Leotiomycetes, Pezizomycetes, Saccharomycotina, and Sordariomycetes. The sequences have been aligned using the software ClustalW (Higgins, D.G., Thompson, J.D., Gibson, T.J., *Methods Enzymol.*, 266, 383–402, 1996). The sequences were extracted from the Genome Portal of the Department of Energy Joint Genome Institute (http://genome.jgi.doe.gov; Nordberg, H., Cantor, M., Dusheyko, S., Hua, S., Poliakov, A., Shabalov, I., Smirnova, T., Grigoriev, I.V., Dubchak, I., *Nucleic Acids Res.*, 42, D26–D31, 2014).

(*Continued*)

an enzyme that directly and specifically oxidizes spermine found (spermine oxidase), whereas in other groups of organisms, the same enzyme is involved in the oxidation of spermidine and spermine. The possibility that the gene encoding this enzyme could have arisen after a duplication event of spermidine oxidase and divergent evolution has been entertained (Polticelli et al. 2012; Cervelli et al. 2013).

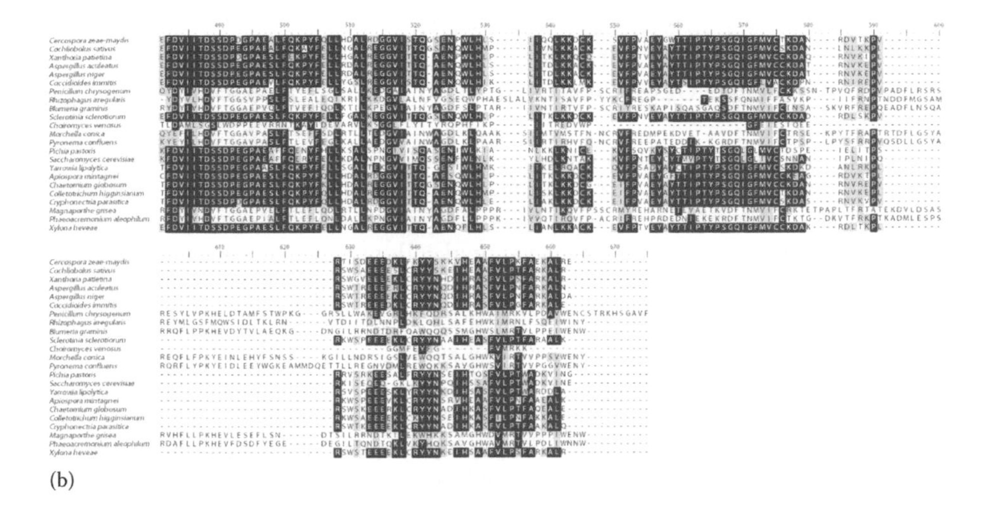

(b)

Figure 4.17 **(Continued)** (a,b) Alignment of the Spm protein sequences of some fungal member belonging to Dothideomycetes, Eurotiomycetes, Glomeramycota, Leotiomycetes, Pezizomycetes, Saccharomycotina, and Sordariomycetes. The sequences have been aligned using the software ClustalW (Higgins, D.G., Thompson, J.D., Gibson, T.J., *Methods Enzymol.*, 266, 383–402, 1996). The sequences were extracted from the Genome Portal of the Department of Energy Joint Genome Institute (http://genome.jgi.doe.gov; Nordberg, H., Cantor, M., Dusheyko, S., Hua, S., Poliakov, A., Shabalov, I., Smirnova, T., Grigoriev, I.V., Dubchak, I., *Nucleic Acids Res.*, 42, D26–D31, 2014).

As mentioned previously, the search of homologous fungal genes described in this chapter revealed the presence of some chimeric products, such as Odc from *A. sarcoides* that contains at its *N*-terminal region homology with a presequence of a protease protein, as well as Odc from *R. allomyces* that is fused to a region with homology with 3-ketoacyl-CoA thiolase at its *N*-terminal region, Samdc from *B. dendrobatis* fused with an F-box domain at its *C*-terminal region, and Basidiomycota Spds, fused to saccharopine dehydrogenase. Other sequences were larger than the average size, such as Odcs from *B. cinerea* and *M. osmundae*; Spdss from *A. fumigatus*, *A. niger*, *P. chrysogenum*, *A. sarcoides*, and *S. sclerotiorum*; and Samdc from *M. globosa* that had no obvious homology with any reported gene. These few examples indicate that events of modification of genes encoding polyamine biosynthetic enzymes in fungi are not rare. Fusion enzymes have been also reported in α, β, and δ proteobacteria, as well as actinobacteria; some of these fusions occurred between adjacent open reading frames, but this was not the most common event. Eukaryotes such as the green alga *Micromonas pusilla* has Odc fused with Samdc, whose origin seems to be the product of an independent event to the one occurring in Basidiomycota (Green et al. 2011). The benefit of gene fusion has been discussed in the context of biosynthetic cost; however, there is not enough information that allows explaining a common or important role for all of these events.

4.5 ADVANTAGES OF FUNGI AS MODELS FOR THE STUDY OF DIFFERENT ASPECTS OF POLYAMINE METABOLISM AND THEIR MODES OF ACTION

The emergence of fungi as model systems for different biological phenomena has grown in importance in the last decades, since fungi offer many advantages over other systems (reviewed by van der Klei and Veenhuis 2006). Regarding polyamine research, it is well known that plant and animal cells present redundancy in the genes encoding the enzymes implicated in polyamine metabolism, a circumstance that may not allow one to answer fundamental questions about these molecules, such as the molecular mechanisms behind their mode of action or the specific functions of each one of the polyamines existing in the cell. Regarding the question of whether each polyamine possesses a specific role in growth and metabolism, little is known about it. We have previously described that the phytopathogenic Basidiomycota *U. maydis* contains only putrescine and spermidine and has a single copy of the genes involved in the synthesis of the two polyamines. Our data demonstrate that putrescine serves only as the biosynthetic precursor of spermidine and that, accordingly, spermidine is the only polyamine absolutely required by the fungus (Valdes-Santiago, Cervantes-Chavez, and Ruiz-Herrera 2009). In organisms containing spermine, the physiological role of this polyamine remains unclear, since, as already indicated previously, it has been demonstrated that this polyamine is not essential for normal growth in fungi; e.g., *S. cerevisiae* Spms mutants grow as well as the wild-type strain in the absence of exogenous polyamines (Hamasaki-Katagiri et al. 1998; Chattopadhyay, Tabor, and Tabor 2003a). In contrast, as indicated previously, spermidine is considered to be essential, probably because it fulfills all the functions attributed to polyamines in general.

Additionally, it has been described (and we discuss it more extensively in Chapter 6) that spermidine serves as a butylamine donor for the synthesis of hypusine present in the eukaryotic initiation factor 5A (eIF5A). eIF5A is a small protein highly conserved and essential in eukaryotes, as demonstrated by gene disruption in *S. cerevisiae* and in mouse embryos (Schnier et al. 1991; Nishimura et al. 2011), and has acquired further importance since its absence is related with apoptosis in different kinds of cancer (Caraglia et al. 2003; Scuoppo et al. 2012; Fujimura et al. 2014); e.g., human cervical presented overexpression of eIF5A (Mémin et al. 2013). Likewise, it has been observed that the inactive form of eIF5A (nonhypusinated) induces cell death in malignant cells (Francis et al. 2014). The increasing interest in eIF5A has resulted in the development of a platform containing structural information of eIF5A (Bertucci Barbosa et al. 2014). The crystal structures of eIF5A have been determined from many organisms, including *S. cerevisiae* (Figure 4.18) (Kim et al. 1998; Yao et al. 2003; Bosch and Hol 2004; Sanches et al. 2008; Tong et al. 2009; Teng et al. 2009).

Considering the importance of spermidine in this essential reaction, the possibility that spermine can substitute for spermidine is unknown. Nevertheless, we embarked ourselves in answering this important question using *U. maydis* as a suitable model since, as already indicated, this Basidiomycota fungus does not contain a Spms gene

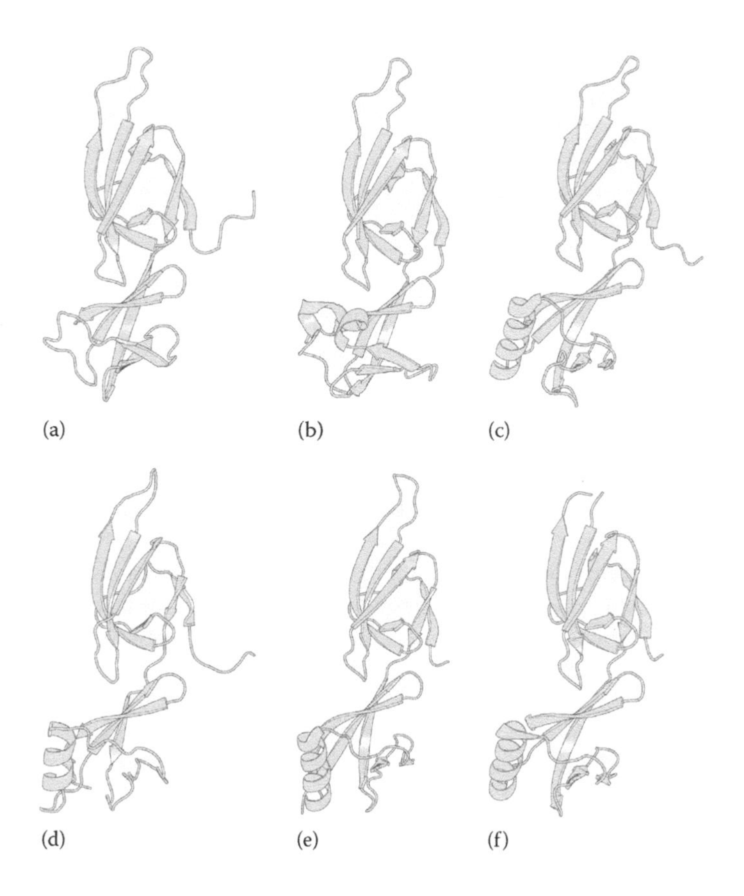

Figure 4.18 Comparison of the structure of the eukaryotic translation initiation factor 5A (eIF5A) from different organisms. (a) The structure of eIF5A from *Methanococcus jannaschii*, 2EIF (Kim, K.K., Hung, L.W., Yakota, H., Kim, R., Kim, S.H., *Proc. Natl. Acad. Sci. USA*, 95, 10419–10424, 1998). (b) Structure of eIF5A from *Pyrococcus horikoshii*, 1ZI6 (Yao, M., Ohsawa, A., Kikukawa, S., Tanaka, I., Kimura, M., *J. Biochem.*, 133, 75–81, 2003). (c) Structural analysis of *Leishmania mexicana*, PDB ID: 1XT. Deukaryotic initiation factor 5a (Bosch, J., Hol, W.G.J., Structural Genomics of Pathogenic Protozoa Consortium, RCSB Protein Data Bank, 2004, http://www.rcsb.org/). (d) Crystal structure of the full-length eIF5A from *Saccharomyces cerevisiae*, PDB ID: 3ER0 (Sanches, M., Dias, C.A.O., Aponi, L.H., Valentini, S.R., Guimaraes, B., Crystal structure of the full length eIF5A from *Saccharomyces cerevisiae* RCSB-Protein Data Bank, 2008, http://www.rcsb.org/). (e) Structure of the eIF5A from *Arabidopsis thaliana*, 3HKS (Teng, Y.B., He, Y.X., Jiang, Y.L., Chen, Y.X., Zhou, C.Z., *Proteins*, 77, 736–740, 2009). (f) Structure of human eIF5A, 3CPF (Tong, Y., Park, I., Hong, B.S., Nedyalkova, L., Tempel, W., Park, H.W. *Proteins*, 75, 1040–1045, 2009). The protein structures were obtained from the RCSB Protein Data Bank, http://www.rcsb.org (Berman, H.M., Westbrook, J., Feng, Z., Gilliland, G., Bhat, T.N., Weissig, H., Shindyalov, I.N., Bourne, P.E., *Nucleic Acids Res.*, 28, 235–242, 2000). This figure was created with Pymol (DeLano, W.L., *The PyMOL Molecular Graphics System*, DeLano Scientific LLC, San Carlos, CA, 2002; http://www.pymol.org).

nor its product, spermine (Valdes-Santiago, Cervantes-Chavez, and Ruiz-Herrera 2009; Valdes-Santiago, Guzman-de-Pena, and Ruiz-Herrera 2010). To determine whether spermine was able to sustain the growth of *U. maydis*, we utilized the strain LV54, an Spds-less mutant (*spe*). The polyamine pool of the strain was depleted by means of growth in the absence of spermidine by several cycles until growth stopped. Afterward, the strain was grown in minimal medium in the presence of 0.1 mM of spermine. During the first 24 hours, the strain was unable to grow, in contrast to the cells grown in a medium containing 0.1 mM spermidine. However, after 48 hours, the cells started to grow in the medium containing spermine only (Valdes-Santiago and Ruiz-Herrera, unpublished data), while in the control without any polyamine added, no growth was observed, indicating that growth in the spermine medium was due to the presence of this polyamine and not to a possible remnant of spermidine present in the cells. At first sight, these results seemed to be in contradiction to the general idea that spermidine is essential for eukaryotic cells and that no other polyamine can fulfill the spermidine roles (Pitkin and Davis 1990; Chattopadhyay, Tabor, and Tabor 2003a; Valdes-Santiago, Cervantes-Chavez, and Ruiz-Herrera 2009). However, the possibility existed that mutants were able to back-convert spermine into spermidine by the action of the polyamine acetylase and oxidase. To eliminate this possibility, we measured the effect of spermine on a mutant unable to synthesize spermidine by both the synthetic and the backward mechanisms. This is a double mutant (*U. maydis odc/pao*) deficient in *ODC* and *PAO* genes. Polyamine-depleted cells were grown in minimal medium in the presence of 5 μM, 0.01, 0.05, 0.5, 1, or 5 mM spermine (Figure 4.19), and the absence of spermidine was confirmed by high-performance liquid chromatography (HPLC). Growth of the double mutant *odc/pao* did not occur in the absence of spermine, and growth with spermine was delayed and reduced in comparison with cell growth with spermidine, but the important issue is that the double mutant grew in the absence of spermidine (Figure 4.19).

Interestingly, we observed that *odc/pao* mutant cells displayed different morphological phenotypes when grown in the presence of spermidine or spermine. Accordingly, cells clearly showed a difference in size and appeared swollen when grown with spermine. Besides, these cells displayed a problem in the separation of mother and daughter cells during budding that could be seen in the form of flower-shaped cell aggregates (Figure 4.20), and this effect was accentuated over time. In contrast, the spermidine-supplemented *odc/pao* mutants showed the normal morphology of the fungus. This difference may be the result of an alteration in the cell cycle brought about by a faulty role of spermidine, suggesting that spermidine may have a role during the cell cycle, specifically during cytokinesis. This observation is not new; in mammalian cells, it was shown that the inhibition of polyamine biosynthesis inhibited cytokinesis and induced the formation of binucleate cells, effects that were reversed by increasing the intracellular levels of the polyamines (Sunkara et al. 1979). Recently, *S. pombe* deleted in the gene encoding Samdc, and which had an absolute requirement of spermidine for growth, showed delay of the cell cycle progression and increase in cell size (Balasundaram, Tabor, and Tabor 1991; Chattopadhyay, Tabor, and Tabor 2002), showing that spermidine plays an important role in the cell cycle in *S. pombe* (Chattopadhyay, Tabor, and Tabor 2002). These

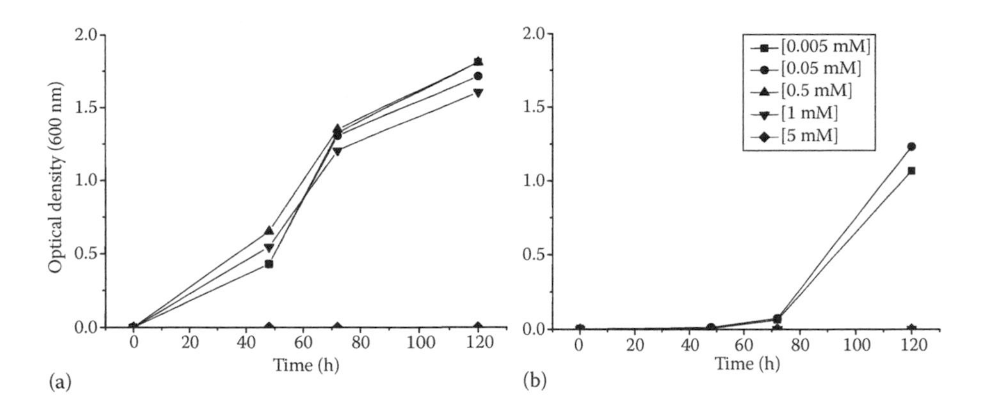

Figure 4.19 Cell growth rate of *odc/pao* double mutant of *Ustilago maydis* incubated with different concentrations of spermidine (a) or spermine (b).

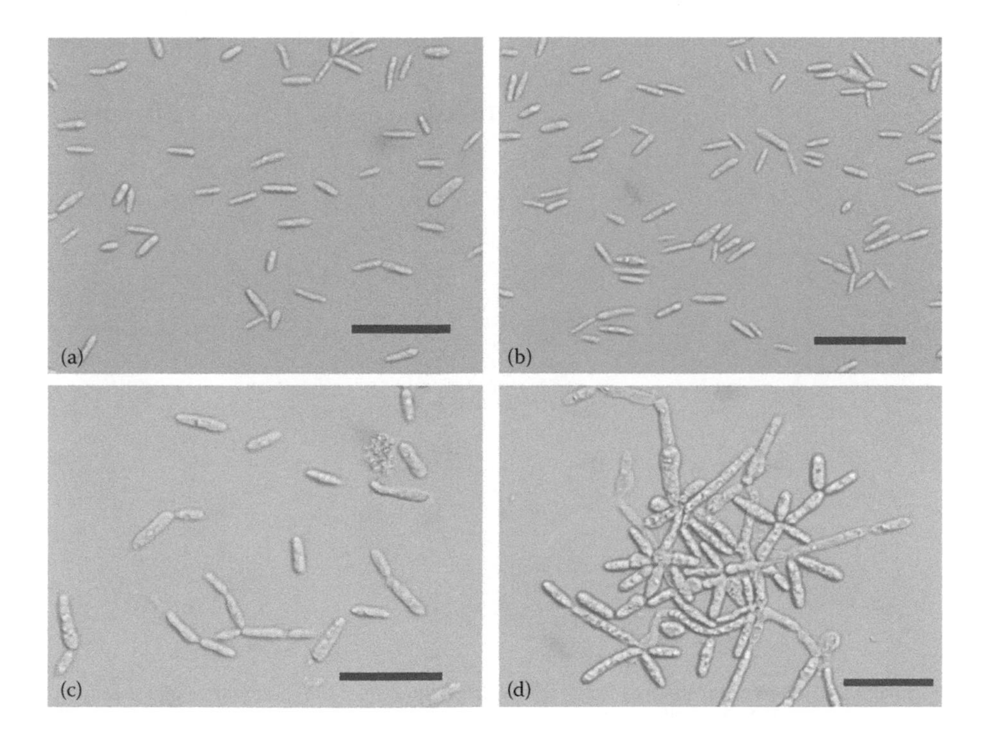

Figure 4.20 Morphological phenotype of *odc/pao* mutant growing with 0.1 mM spermidine (a,b) or 0.1 mM spermine (c,d). Mutants were grown at 28°C at different incubation periods: a and c, 36 hours; b and d, 64 hours. Magnification bar, 30 μm.

results would represent the novel finding that spermine may substitute for spermidine in some of the essential roles of polyamines, but with low efficiency, and may be relevant to explain, perhaps as a safety mechanism, the anomalous phenomenon that higher eukaryotic organisms contain spermine, no matter that spermidine would be sufficient to satisfy their polyamine requirements.

However, another possible explanation is possible. The most serious implication of the substitution of spermine by spermidine, as it was mentioned earlier, is that spermidine is a substrate for the synthesis of hypusine, a needed modification to the activation of the essential protein eIF5A. Hypusine is formed from spermidine by the sequential action of two enzymes; thus, hypusine biosynthesis starts with the activity of deoxyhypusine synthase (Dhs; EC 1.1.1.249). DHS transfers a 4-aminobutyl moiety from spermidine to the epsilon-amino group of a conserved lysine residue within the inactive eIF5A precursor protein chain (Figure 4.21) (Park, Wolff, and Folk 1993a; Park, Lee, and Joe 1997). The resultant deoxyhypusine intermediate is further converted to the active form that is essential for eukaryotic cell growth and proliferation (Schnier et al. 1991; Sasaki, Abid, and Miyazaki 1996; Park 2006). Accordingly, it is possible that Dhs from *U. maydis* recognizes both spermidine and spermine as substrate, in such a way that activation of eIF5A is possible and does not affect the viability of the double mutants *odc/pao* when they grow with spermine only, or else that modification of eIF5A by deoxyhypusine in some fungi is important, but not strictly necessary. In this regard, it is important to note that an *in silico*

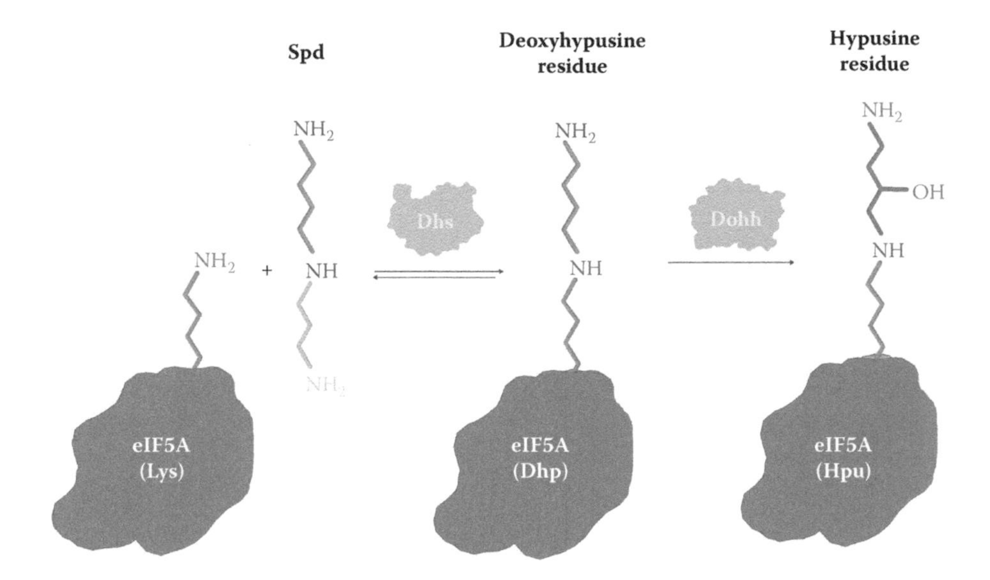

Figure 4.21 **(See color insert.)** Schematic representation of the mechanism of activation of eIF5A. A modification in a specific lysine residue (blue) on eIF5A is performed by the transfer of an 4-aminobutyl moiety (red) from spermidine to lysine ε-amino group yielding Dhp, which is hydroxylated by the action of Dohh to form Hpu. Dhp, deoxyhypusine; Dhs, deoxyhypusine synthase; Dohh, deoxyhypusine hydrolase; Hpu, hypusine; Lys, lysine; Spd, spermidine.

analyses revealed the existence of a single-copy homologous sequence encoding Dhs in *U. maydis* with number um00071.

To be conclusive in determining the more adequate response to this problem, it would be necessary to clone and express the putative *DHS* gene from *U. maydis* to analyze its specificity toward spermine and spermidine. It may be indicated that Dhs activity assay measuring the incorporation of aminobutyl moiety coming from [^{14}C]-spermidine has been reported previously (Ober and Hartmann 1999; Njuguna et al. 2006; Wolff, Lee, and Park 2011).

Another further possibility to explain our results is that before being recognized as a substrate, spermine suffers some modifications. There is evidence that some analogs of spermidine such as α- and β-methylated spermidines were effective in supporting long-term growth and serve as substrates for deoxyhypusine synthesis of polyamine depleted DU145 cells (human prostate carcinoma cell line) treated with an inhibitor of polyamine biosynthesis (Hyvonen et al. 2011). Thus, the possibility existed that spermine suffered some modifications when used as a sole source of polyamines in an *odc*/*pao* double mutant that led to the formation of some active spermidine derivatives. To address this possibility, we proceeded to separate by chromatography (see previous discussion) the polyamines synthesized by the double mutant and analyzed all the products obtained by liquid chromatography tandem-mass spectrometry (LC-MS/MS). The results of our MS analyses revealed the presence of peaks with m/z that corresponded to the following spermine derivatives: acetylated spermine, 452.6; dimethylspermine, 454.66; diethylspermine, 362.56; and ethylspermine, 438.62 (Valdes-Santiago and Ruiz-Herrera, unpublished data). These results do not support the hypothesis that spermine were transformed into spermidine derivatives, giving more credence to the possibility that spermine may substitute for spermidine in the synthesis of a derivative of hypusine.

CHAPTER 5

Regulatory Mechanisms of Polyamine Metabolism at the Transcriptional, Translational, and Posttranslational Levels in Fungi and Other Organisms

5.1 INTRODUCTION

It is possible to state that the mechanisms involved in polyamine regulation probably emerged to avoid the accumulation or depletion to dangerous levels of intracellular polyamines, as well as to provide enough polyamine concentrations in time and space during the occurrence of specific processes such as cell proliferation and differentiation, as well as during stressful conditions where higher levels of polyamines are required. Accordingly, it is known that an excess of polyamines can be relieved through their sequestration within vacuoles or even through catabolization, since while polyamines are essential, at the same time, high levels may become cytotoxic. Other mechanisms for the regulation of polyamine concentration take place at the levels of transcription, translation, and posttranslation processes. The interplay of all these mechanisms secures that the levels of polyamines are tightly regulated by a quite complex mechanism, as discussed in this chapter.

Polyamine homeostasis depends, as indicated, on the biosynthesis and degradation processes of the enzymes involved in their metabolism, mainly ornithine decarboxylase (Odc) and *S*-adenosylmethionine decarboxylase (Samdc). The mechanism involved in the regulation of Odc is well known because of its short half-life (Russell and Snyder 1969; Pegg 1986). This mechanism of regulation is very interesting. In contrast with proteins that are normally targeted by ubiquitination to be degraded by the 26S proteasome, Odc is targeted without ubiquitination with the help of a small protein named antizyme (Az). The levels of Az are, in turn, regulated by both a programmed frame shifting of the Az mRNA and by an Az inhibitor (AzI).

Regarding Samdc, it is synthesized as a proenzyme; after autocatalytic cleavage at a specific serine residue, it generates two distinct subunits that serve to produce the mature Samdc. Regarding the other aspect of the process, we know that spermidine/spermine-N^1-acetyltransferase (Ssat) is essential for polyamine catabolism. Thus, the products of degradation of polyamines, such as N^1-acetyl polyamines generated by the Ssat, are exported out of the cell, allowing polyamine pools to be maintained in levels that sustain cell growth and allow the response to different stimuli. In this way, polyamine catabolism also contributes to polyamine homeostasis. Interestingly, if the cells require an increase in the levels of putrescine or spermidine, N^1-acetyl polyamines are not exported out of the cell and rather they are oxidized by polyamine oxidase (Pao) to produce either spermidine from spermine or putrescine from spermidine.

It has been reported that the control of all these enzymes at different levels, either at the transcriptional or translational level, in fungi is similar to the mechanisms found in higher organisms.

5.2 REGULATION OF THE SYNTHESIS OF ODC AT THE TRANSCRIPTIONAL AND POSTTRANSCRIPTIONAL LEVELS

Transcriptional and posttranscriptional mechanisms that are involved in the control of Odc synthesis were originally discovered in mammalian cells. Thus, it was found that the levels of Odc biosynthesis can be regulated by factor called 4E-BP1 (4E-binding protein) that binds to the eukaryotic translation initiation factor eIF4E. When the level of the complex eIF4E and 4E-BP1 is high, Odc translation is decreased, and through this mechanism, polyamines could be regulating their own biosynthesis (Flynn 1996; Graff et al. 1997). Translational regulation of Odc depends largely on the 5′ and 3′UTRs of the *ODC* mRNA. In Odc from mammalian reticulocyte cells, it was found that low levels of spermidine induced its synthesis, whereas high levels of spermidine inhibited its synthesis. It is known that variation in the 5′UTR size affects translational efficiency, and in this system, the region –70 to 220 was found to be important to repress transcription by spermidine addition and the region –70 to 170 was involved in stimulation of transcription by low levels of spermidine (Ito et al. 1990). A more detailed analysis showed that the region –70 to 170 contained a GC-rich sequence important for the regulation by spermidine (Kashiwagi, Ito, and Igarashi 1991). In *Saccharomyces cerevisiae*, immunoprecipitation studies showed that Odc levels were the same in cell-free extracts of the wild-type strain compared with *S. cerevisiae samdc* mutants; however, in *samdc* mutants, Odc activity was almost absent, suggesting a posttranslational modification that induced the loss of the enzymatic activity (Fonzi 1989).

Using *S. cerevisiae* as a model, it has been demonstrated that unlike mammalian Odc, an *N*-terminal domain of the Odc is essential for its degradation; however, no specific sequence was required, since cells where there occurred replacement of the native *N*-terminal domain by the *Escherichia coli* lacI repressor behaved exactly like the wild-type Odc (Godderz et al. 2011). The evidences suggested that the unstructured domain in an appropriate context alone is sufficient to target the enzyme for ubiquitin-independent degradation by the proteasome, and this notion could explain the high degree of dissimilarity in this domain of *ODC* genes from related yeast species (Godderz et al. 2011). In addition, the rate of Odc degradation in *S. cerevisiae* is increased by the addition of spermidine to the culture medium. This behavior suggested an Az-like mechanism for the regulation of Odc levels (Gupta et al. 2001; Palanimurugan et al. 2004). In the case of *Neurospora crassa*, two regions of the *ODC* transcript that affect its expression were identified. The elimination of one of them, an upstream activation sequence, reduced mRNA abundance fivefold, together with the loss in the regulation by polyamines, while the other one was a segment of the 5′UTR, whose deletion induced the expression of the *ODC* gene and decreased its regulation by polyamines (Pitkin et al. 1994). Interestingly, the regulation of *ODC* in *N. crassa* requires complex interactions between the 5′UTR and the 3′activator sequence elements and upstream regions (Hoyt, Broun, and Davis 2000). Posttranscriptional regulation of *ODC* was suggested in *Candida albicans* during differentiation, since higher levels of Odc activity occurred, with no increase in its mRNA levels (López et al. 1997).

5.3 THE UNIQUE MECHANISM OF AZ AS A REGULATORY PROTEIN

Odc activity regulation is performed principally by a unique mechanism that involves the degradation of the protein by the 26 proteasome through an ubiquitin-independent manner (Murakami et al. 1992; Hayashi, Murakami, and Matsufuji 1996; Coffino 2001a). When the intracellular concentration of polyamines increases, the synthesis of a protein designated as Az, whose structure is shown in Figure 5.1 (Matsufuji et al. 1995), is induced. In turn, Az interacts with the enzymatically active homodimer (Odc/Odc) complex to form an inactive heterodimer (Odc/Az) (Figure 5.2). This complex is then recognized and degraded by the proteasome (Li and Coffino 1992; Elias et al. 1995; Ivanov et al. 1998; Ivanov, Gesteland, and Atkins 2000). Az was originally recognized by its intervention in Odc degradation, but further on, an additional role in polyamine transport has been suggested (Mitchell et al. 1994; Suzuki et al. 1994; Hayashi, Murakami, and Matsufuji 1996; Coffino 2001b).

All known Azs are encoded by two overlapping open reading frames (ORF1 and ORF2); ORF1 is very short and contains a start codon and a stop codon, while ORF2 does not contain an obvious start codon but encodes the greater part of the region encoding the protein. In consequence, a frame shift in the +1 reading frame at the internal stop codon is necessary to form the complete and functional Az (see a schematic representation in Figure 5.3) (Rom and Kahana 1994; Matsufuji et al. 1995). In turn, Az synthesis and degradation are regulated by the polyamine levels of the cells: an elevation in polyamine concentration prevents Az degradation and at the same time induces Az synthesis. Az degradation is ubiquitin dependent (Gandre, Bercovich, and Kahana 2002; Palanimurugan et al. 2004). *S. cerevisiae* contains only one Az orthologue (Oaz1). Different versions of Oaz1 with truncations at the 5′ end of the coding sequence showed inhibition of Oaz1 translation, whereas complete inhibition occurred when the 3′ end was deleted. This result suggested that both extremes of the protein are required for polyamine sensing; however, silent mutations had no effect on the +1 ribosomal frame shift efficiency, indicating that the nascent polypeptide was required for the regulation. This hypothesis was confirmed with the observation that at low polyamine levels, nascent Oaz1 polypeptide piles up

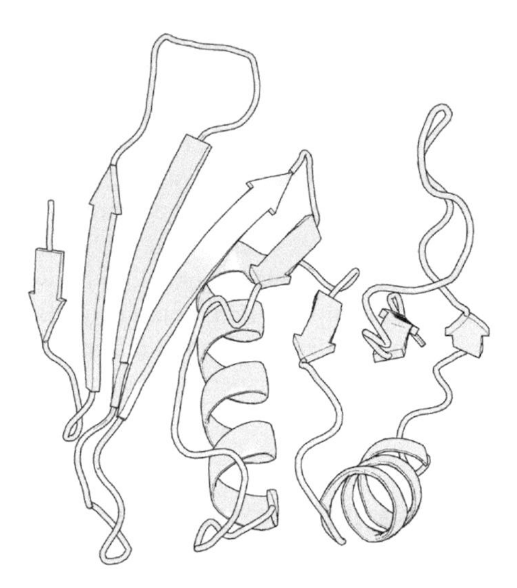

Figure 5.1 Structure of the antizyme (Az) protein, 3BTN (Albeck et al. 2008), created with Pymol (DeLano, W.L., *The PyMOL Molecular Graphics System*, DeLano Scientific LLC, San Carlos, CA, 2002; http://www.pymol.org).

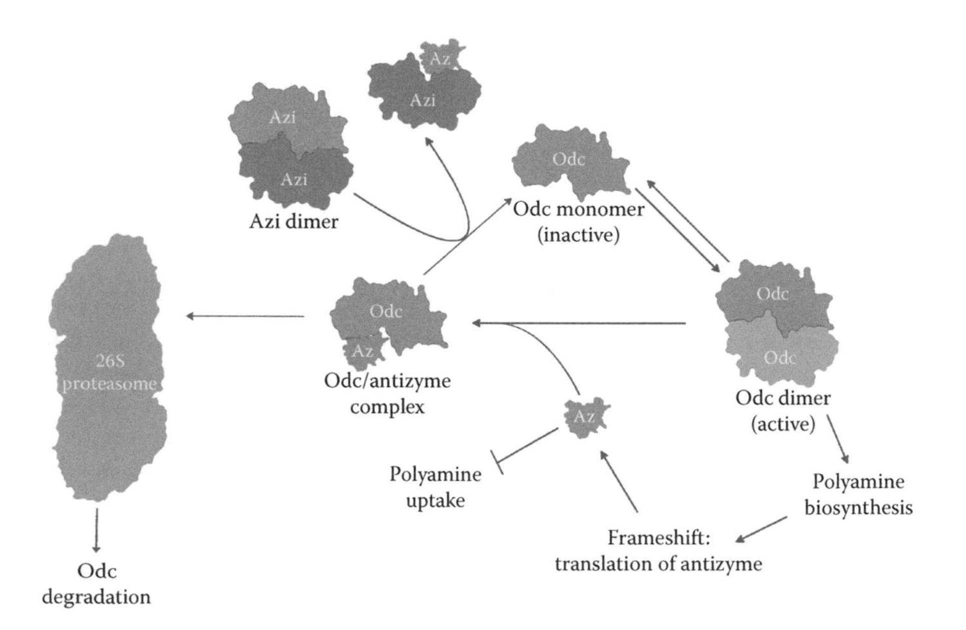

Figure 5.2 **(See color insert.)** Schematic representation of the ornithine decarboxylase (Odc) regulation mediated by antizyme (Az). Odc homodimer is the active form. Az inhibits the formation of the Odc homodimer, leading Odc for degradation via the 26S proteasome without ubiquitination. At the same time, Az inhibits polyamine uptake. Polyamine biosynthesis is restored by displacement of Az by the Az inhibitor (AzI), allowing Odc homodimer formation.

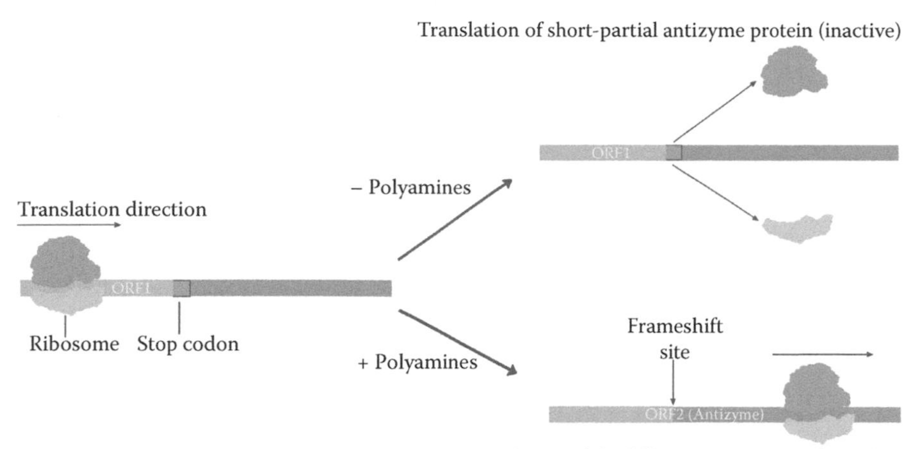

Figure 5.3 **(See color insert.)** Schematic representation of the regulation of antizyme (Az) synthesis by polyamines. Az mRNA contains two overlapping open reading frames (ORFs). ORF1 contains a UGA stop codon that is read under conditions of low levels of polyamines, resulting in no functional Az. High levels of polyamines stimulate a +1 frame shift, allowing continuation of translation of ORF2, resulting in the synthesis of the complete and active Az form.

and is found associated with ribosomes just at the ribomal frame shift site, while at high polyamine levels, there occurs an impediment to achieve the inhibitory conformation, allowing the translation of the functional Oaz1 to occur (Palanimurugan et al. 2004; Kurian et al. 2011).

In mammalian cells, the Az family is made up of at least four members, and Azs have their own inhibitor, a protein termed AzI (Figure 5.4). Interestingly, at the structural level, AzI contains high homology with Odc, but of course, it is devoid of any decarboxylase activity (Murakami et al. 1989; Coffino 2001b; Mangold and Leberer 2005; Mangold 2006). It has been described that mammalian cells contain at least two AzIs, and by use of yeast three-hybrid assays, it was determined that AzI impedes the interaction of Odc with Az (Mangold and Leberer 2005).

The presence of a homolog of the inhibitor of Az1 has not been described in fungi, but our *in silico* blast analyses demonstrate the presence of homologs of the gene encoding Az in a least several races of *Fusarium oxysporum* (Valdés-Santiago, unpublished observations).

A detailed analysis has shown the absence of Az homologues in plants, but an Az gene has been found in the photosynthetic organism *Euglena gracilis* (Ivanov and Atkins 2007). Accordingly, it is possible that plants once contained Az genes and that they were lost in one of their common ancestors. A further explanation to this phenomenon has been advanced (Ivanov and Atkins 2007). These authors pointed out the relatively low homology among Azs from different sources and hypothesized that plant Azs have evolved in such a way that their identification by sequence comparisons to other Azs is not possible. In plant

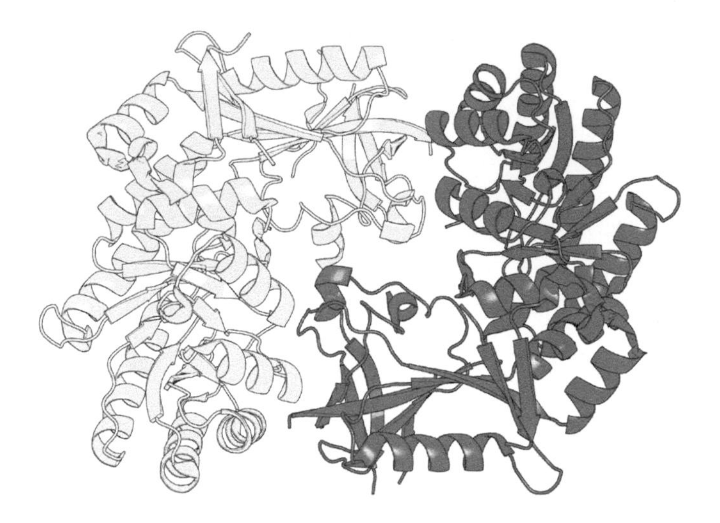

Figure 5.4 Structure of the homodimer of the antizyme inhibitor (AzI) protein, 3BTN (Albeck et al. 2008) created with Pymol (DeLano, W.L., *The PyMOL Molecular Graphics System*, DeLano Scientific LLC, San Carlos, CA, 2002; http://www.pymol.org). Each monomer is highlighted with different levels of gray scale.

parasitic organisms such as *Phytomonas* spp., the absence of Azs has been proposed, and in other parasites, such as *Crithidia fasciculata*, it was found that Odc was degraded by poly-ubiquitination of the enzyme (Persson, Jeppsson, and Nasizadeh 2003; Marcora et al. 2010). Nevertheless, a putative inhibitor of Odc was reported in barley seeds, a protein of ≈16 kDa that blocked Odc activity (Kyriakidis 1983). By use of yeast two-hybrid assays, a cytosolic ribosomal plant Odc binding protein (rpS15) was isolated. Nevertheless, no posttranscriptional regulation of Odc or arginine decarboxylase (Adc) has been proved, and rpS15 did not present interaction either with human or *S. cerevisiae* Odcs and did not reduce Odc activity. All these data constitute strong evidence that regulation of the intracellular concentrations of polyamines in plants seems not to be related with Az (Illingworth and Michael 2012).

In those fungal systems that have been analyzed, they were found to posses only a single copy of the gene encoding Az (Ivanov et al. 2000; Palanimurugan et al. 2004; Ivanov, Gesteland, and Atkins 2006). In *Schizosaccharomyces pombe*, Az is not essential, since growth of mutants affected in the gene encoding Az is not affected, although mutants presented a much higher level of polyamines (40-fold) than the wild-type strain did (Ivanov et al. 2000). In the case of *S. cerevisiae* Az, it was found that it binds to Odc in a 1:1 relationship and leads to the degradation of the yeast but not the human Odc by the proteasome (Porat et al. 2008; Godderz et al. 2011; Chattopadhyay et al. 2011a). Yeast Az is a valuable system to understand Az functions since it presents similar features to those from Azs from mammalian cells, such as translational frame shifting and the general mechanism by which degradation of Odc takes place (Li and Coffino 1992; Palanimurugan et al. 2004;

Chattopadhyay et al. 2011a,b), and as described previously, fungi contain a single gene encoding this inhibitory protein.

Also in *S. cerevisiae*, it was described that eRF3, a GTPase that induces the release of the nascent polypeptide chain stimulating the translation termination process, is converted in the aggregated conformation of a prion form named PSI+ (Salas-Marco and Bedwell 2004; Urakov et al. 2006). Interestingly, PSI+ incremented the expression of *AZ*, inducing in this way the degradation of Odc and in turn decreasing polyamine levels (Namy et al. 2008). Additionally, Az has been related with the regulation of the cyclin D1 and Smad1, proteins implicated in the cell cycle, and in the function of Mps, implicated in amplification of the centrosomes (Gruendler et al. 2001; Newman et al. 2004; Kasbek, Yang, and Fisk 2010).

5.4 REGULATION OF THE BIOSYNTHESIS OF SAMDC

Mammalian and plant cells contain several copies of the gene encoding Samdc (Pulkka et al. 1991; Maric, Crozat, and Jänne 1992; Sakata et al. 1993; Ge et al. 2006). Interestingly, in opposition with plant and animal cells, the *SAMDC* genes in lower eukaryotes like fungi, as well as in some reported nematodes, exists as a unique copy (Kashiwagi et al. 1990a; Da'Dara and Walter 1998; Valdes-Santiago et al. 2012b). Samdc is regulated at different levels: transcriptional, translational, and posttranslational (Heby and Persson 1990; Pegg and McCann 1992; Pegg et al. 1998; Shantz and Pegg 1999). Samdc is synthesized as a proenzyme, and in mammalian cells, the presence of putrescine is essential for the processing of the proenzyme and also for enzyme activity (Stanley, Pegg, and Holm 1989; Coleman et al. 1994; Pegg et al. 1998). In contrast, *Mucor rouxii* and potato Samdc proenzymes do not require putrescine for processing and activation, while, on the other hand, the yeast enzyme does (Kameji and Pegg 1987; Calvo-Mendez and Ruiz-Herrera 1991; Xiong et al. 1997). In addition to the phenomenon of enzyme maturation, spermidine intracellular concentrations are regulated at other levels. Thus, it has been shown that spermidine depletion provokes an increase in the levels of *Samdc* mRNA and *Samdc* translation, as well as an increase in Samdc activity in animal cells (White et al. 1990; Pulkka et al. 1991; Shantz et al. 1992).

There exist some reported important differences in the regulation of fungal, mammalian, and plant *Samdc* genes; e.g., in plants, Samdc seems to possess predominance in the regulation of polyamine biosynthesis (Illingworth and Michael 2012), and mammals and plants contain an upstream ORF (uORF) that within its leader sequence encodes a peptide named MAGDIS (Met-Ala-Gly-Asp-Ile-Ser). It is known that the number of ribosomes loaded on individual mRNA molecules is controlled by the presence of this upstream uORF in the corresponding messenger (Mach et al. 1986; White et al. 1990; Morris and Geballe 2000; Mize and Morris 2001; Meijer and Thomas 2002; Hu, Gong, and Pua 2005; Tran, Schultz, and Baumann 2008). The uORF down-regulates the translation of the main ORF, and high levels of polyamines inhibit the synthesis of the MAGDIS peptide through ribosome stalling (Ruan et al. 1996).

In contrast to those results, *S. cerevisiae* does not have such uORF (Mize and Morris 2001; Hu, Gong, and Pua 2005; Tran, Schultz, and Baumann 2008), and in fact, there are no reports of uORF presence in any fungal phyla (Mize et al. 1998; Mize and Morris 2001). Furthermore, when we made an *in silico* search in *Ustilago maydis* Samdc promoter using the nucleic acid sequence of the uORF (ATG GCC GGC GAC ATT AGC TAG) and the translated sequence, we found no coincidences (Valdes-Santiago, unpublished data). This result confirms that the previously mentioned regulation mechanism is not conserved in the fungal kingdom.

5.5 REGULATION OF THE ACTIVITY OF THE POLYAMINE BIOSYNTHETIC ENZYMES

The activity of enzymes is the sum of several factors such as transcription of its mRNA, synthesis, degradation, and substrate availability. As described already, polyamine biosynthetic enzymes are characterized by their rapid turnover or short half-lives. Thus, cycloheximide addition resulted in a rapid decay in Odc activity in *M. rouxii* (Martínez and Ruiz-Herrera, unpublished observations). Similarly, in mammalian cells, after the addition of cycloheximide to inhibit protein synthesis, it has been observed that Odc activity decays in a period from 5 to 48 minutes depending on the tissues and developmental stage (Russell and Snyder 1969; Iwami et al. 1990; Berntsson, Alm, and Oredsson 1999). Similarly, it is also known that Odc half-life varies depending on the phase of differentiation and whether the cells are in proliferating stages or not. In IEC-6 cells, the Odc half-life was increased from 29 to 35 minutes when cells were stimulated by asparagine; furthermore, it was observed that androgens, such as testosterone, prolonged the half-life of Odc (Laitinen, Laitinen, and Pajunen 1984; Iwami et al. 1990; Sparapani et al. 1998).

The Samdc half-life has been determined to be almost 1 hour (Russell and Snyder 1969; Shirahata and Pegg 1985), and *Trypanosoma brucei* and *Leishmania donovani* Odc and Samdc enzymes presented higher stability compared with fungal or mammalian Odcs (Phillips, Coffino, and Wang 1987; Ghoda et al. 1990; Persson, Jeppsson, and Nasizadeh 2003). However, in *C. fasciculata*, Odc half-life was as short as 3 minutes.

Odc activity is influenced by polyamine levels. *S. cerevisiae* presents this distinctive response, as observed that addition of spermidine or spermine inhibited almost 100% of Odc activity, while putrescine addition affected only slightly the enzyme half-life of the biosynthetic enzymes. The same response has been observed in mammalian cells (Fonzi 1989; Ginty et al. 1990). It seems important to recall that regulation of Odc activity has been observed during the different stages of the cell cycle in yeast, plant, and animal cells (Kay, Singer, and Johnson 1980; Heby 1981; Chang and Chen 1988; Bettuzzi et al. 1999; Kwak and Lee 2002), but as mentioned previously, the mechanism involved in the phenomenon may be different, depending on the organism.

As indicated previously, regulation of Odc by the addition of spermidine or spermine, but not by putrescine, has been reported in *S. cerevisiae*, and Odc activity is almost absent in the first hours after polyamine addition. This loss of activity was explained in terms of protein modification rather than protein degradation, since there was no change in the Odc levels (Tyagi, Tabor, and Tabor 1981). Evidences of translational regulation of Odc by putrescine, spermidine, and spermine have been also reported in mammalian cells (Kahana and Nathans 1985; Persson, Holm, and Heby 1986; Pegg et al. 1988).

Rowland, Morris, and Coffino (1992) have suggested that the absence of evidence of allosteric regulation of Odc and the dramatic changes in polyamine levels without alteration in the growth rate of the organisms, especially in *N. crassa*, could be related to the availability of polyamines and Odc present in the vacuoles, and in this sense, it is opportune to recall the existence of several polyamine pools. Thus, *M. rouxii* Odc presented an interesting subcellular distribution, indicating the existence of more than one Odc pool as described by Martinez-Pacheco and Ruiz-Herrera (1993). In transgenic mouse lines overexpressing Odc, the levels of spermidine and spermine were normal, while pools of putrescine were highly increased. These pools would provide the enzyme and substrate to face with cell necessities (Halmekyto et al. 1991; Rowland, Morris, and Coffino 1992).

Regulation of polyamine biosynthesis may be indirectly affected by the production of metabolites related to polyamine metabolism such as 5′-methylthioadenosine (MTA). MTA is a byproduct of polyamine synthesis, since it is generated from decarboxylated *S*-adenosylmethionine (dcSAM) during the synthesis of spermidine and spermine. MTA is converted into adenine and methylthioribose 1-phosphate by the action of MTA phosphorylase (Mtap). After a series of reactions, both products are converted to adenine nucleotides and methionine (Figure 3.11) (Backlund and Smith 1981; Kamatani and Carson 1981; Cone et al. 1982; Savarese et al. 1983; Marchitto and Ferro 1985). It was observed that the concentration of MTA regulates Odc and Samdc activities in enzyme-deficient lymphoma cells, provoking an increase in putrescine and dcSAM concentration (Kubota, Kajander, and Carson 1985). *S. cerevisiae* contains one *MTAP* gene homologue, and mutants defective in Mtap had a significant increase in Odc activity that leads to the elevation of polyamine concentrations and an increase in Odc activity (Subhi et al. 2003). In addition, it has been suggested that the accumulation of MTA in *S. cerevisiae mtap* mutants inhibits spermidine synthase (Chattopadhyay, Tabor, and Tabor 2006b). A further example regarding these indirect effects is the observation that *in vitro* liver and mammary gland Samdc is repressed by spermine, but not by spermidine (Sakai et al. 1979).

5.6 REGULATION OF POLYAMINES DURING STRESS CONDITIONS: A COMPARATIVE STUDY BETWEEN FUNGAL AND PLANT CELLS

Vast available evidence supports the role of polyamines in the regulation of biological processes, such as increased growth, tissue regeneration, and cellular differentiation, and even virulence and pathogenicity in fungi (Jelsbak et al. 2012; Valdes-Santiago et al. 2012c; Goforth, Walter, and Karatan 2013; Kummasook et al. 2013). Additionally, both fungal and plant cells face normally environmental stresses, and it has been amply demonstrated that polyamines also have a role in stress response (Ahuja et al. 2010; Alcazar et al. 2010; Gupta, Dey, and Gupta 2013; Valdes-Santiago and Ruiz-Herrera 2014; Deryng et al. 2014).

Conservation of signaling pathways related with stress response is prevalent among fungi (Bahn et al. 2007; Nikolaou et al. 2009). In these organisms, evidences show a direct effect rather than a side effect of stress on polyamine metabolism. Accordingly, *U. maydis* impaired in the synthesis of spermidine synthase was significantly more sensitive to treatment with KCl or sodium dodecyl sulfate (SDS), compared with wild-type strain, and the same results were observed with double mutants affected in spermidine synthase (*spds*) and Odc (*odc*) (Valdes-Santiago, Cervantes-Chavez, and Ruiz-Herrera 2009). Something interesting happened with *U. maydis* Pao (*pao*) mutants, and double mutants affected also in Odc (*odc/pao*), under stress induced by NaCl, LiCl, or sorbitol: only the double mutants presented a sensitive phenotype, while *pao* single mutants behaved as the wild-type strain (Valdes-Santiago, Guzman-de-Pena, and Ruiz-Herrera 2010). Since *U. maydis odc/pao* mutants were unable to produce putrescine, it is possible to conclude that spermidine is the polyamine directly involved in response to the tested stress conditions. In this sense, a comparative analysis between Samdc (*samdc*) and *spds* mutants is illustrative. Both enzymes are required for spermidine synthesis, but the mutants showed different responses to the same stress condition induced by LiCl, although, in contrast, their response to high NaCl was similar. This result was explained on the basis of their differences in the accumulation of *S*-adenosylmethionine and dcSAM and its possible effect on DNA methylation (Valdes-Santiago et al. 2012b; Valdes-Santiago and Ruiz-Herrera 2014).

Polyamine uptake seems to be related with osmotic stress provoked by high levels of NaCl, KCl, or sorbitol, since the expression of the *AGP2* gene involved in polyamine import is increased during such stress (Lee et al. 2002; Aouida et al. 2005). Furthermore, spermine uptake is regulated by serine/threonine protein kinases Ptk1p and Ptk2, and disruption in Ptk2 provoked salt tolerance, whereas Ptk2 as well as Sky1p (another serine/threonine kinase) overexpression provoked salt sensitivity in *S. cerevisiae* (Erez and Kahana 2001). An evidence that spermidine could be related with modulation of genes involved with yeast sporulation (a phenomenon that may be related to stressful conditions) is that spermidine synthase is required for repression of NER^{DIT}, a negative regulatory element of sporulation, and the absence of spermidine did not only allow NER^{DIT} to repress completely its target genes but also affected in a moderate way the expression of other genes (Friesen, Hepworth, and Segall 1997; Friesen, Tanny, and Segall 1998).

Oxidative stress might also affect polyamine metabolism, since yeast mutants affected in the production of spermidine or spermine presented diminished cell viability under oxygen toxicity. It was speculated that polyamines might be protecting fungal cells, avoiding the oxidative damage of lipids in the cell and mitochondrial membranes (Balasundaram, Tabor, and Tabor 1993; Chattopadhyay, Tabor, and Tabor 2006a). Accordingly, it is also accepted that another polyamine function is the protection of cells caused by reactive oxygen species (ROS), and in fact, the accumulation of ROS in yeast polyamine-deficient mutants was shown to induce an apoptotic phenotype (Chattopadhyay, Tabor, and Tabor 2006a; Rider et al. 2007; Cerrada-Gimenez et al. 2011). Nevertheless, in plants, ROS generated during both polyamine oxidation and retroconversion has been related with response to different conditions and plant defense (Moschou and Roubelakis-Angelakis 2011; Pottosin et al. 2012).

Another stress related with polyamines is resistance to high temperatures. Thus, *odc* null mutants of *Tapesia yallundae, U. maydis,* and *S. cerevisiae* showed sensitivity to high temperatures comparatively with wild-type strains (Mueller et al. 2001; Balasundaram, Tabor, and Tabor 1996; Cheng et al. 2009; Valdes-Santiago and Ruiz-Herrera 2014). In the plant model *Arabidopsis thaliana,* spermine overexpression induced resistance to high temperatures, and cold conditions resulted in the increase in putrescine levels, as well as in the expression of genes encoding Adc and Samdc (Urano et al. 2003; Cuevas et al. 2008; Bibi, Oosterhuis, and Gonias 2010; Sagor et al. 2013).

In the case of plants, polyamine metabolism has been recommended as a perfect target to improve stress tolerance (Minocha, Majumdar, and Minocha 2014; Shi and Chan 2014). According to different reports, changes in polyamine concentration take place as a response to stress conditions in the form of an adaptive mechanism. Examples of plants where osmotic shock induces an increase in putrescine, spermidine, and spermine levels are wheat, barley, corn, wild oat leaves, and *A. thaliana* (Flores and Galston 1982; Alet et al. 2012; Grzesiak et al. 2013). In opposition, potato and *Bromus* presented a decrease in putrescine, spermidine, and spermine under osmotic conditions (Gicquiaud, Hennion, and Esnault 2002; Liu et al. 2005). Several other contradictory results proving that it is not possible to establish a common pattern, indicating that higher levels of polyamines correspond to higher level of stress protection, have been thoroughly reviewed by Pottosin and Shabala (2014). Nevertheless, osmotic stress resistance was observed in transgenic plants such as *A. thaliana* expressing the35S::*Poncirus trifoliate ADC* gene and *Nicotiana tabacum* expressing the 35S::*Malus domestica SAMDC2* gene (Zhao et al. 2010; Wang et al. 2011). Other examples of stress resistance induced by the overexpression of genes involved in polyamine metabolism are reviewed elsewhere (Gicquiaud, Hennion, and Esnault 2002; Minocha, Majumdar, and Minocha 2014). Although expression of multicopy polyamine metabolic genes in plants did not allow a confirmatory analysis of the relation of a determined phenotype with stress response, *A. thaliana* mutants affected in the production of spermine showed hypersensitivity to ionic stress produced by NaCl or KCl (Yamaguchi et al. 2006; Janicka-Russak et al. 2010).

Polyamine uptake seems to be related with osmotic stress provoked by high levels of NaCl, KCl, or sorbitol, since the expression of the *AGP2* gene involved in polyamine import is increased during such stress (Lee et al. 2002; Aouida et al. 2005). Furthermore, spermine uptake is regulated by serine/threonine protein kinases Ptk1p and Ptk2, and disruption in Ptk2 provoked salt tolerance, whereas Ptk2 as well as Sky1p (another serine/threonine kinase) overexpression provoked salt sensitivity in *S. cerevisiae* (Erez and Kahana 2001). An evidence that spermidine could be related with modulation of genes involved with yeast sporulation (a phenomenon that may be related to stressful conditions) is that spermidine synthase is required for repression of NER^{DIT}, a negative regulatory element of sporulation, and the absence of spermidine did not only allow NER^{DIT} to repress completely its target genes but also affected in a moderate way the expression of other genes (Friesen, Hepworth, and Segall 1997; Friesen, Tanny, and Segall 1998).

CHAPTER 6

General Functions of Polyamines and Their Possible Modes of Action in Fungi and Other Organisms

6.1 INTRODUCTION

The field of polyamine research has blossomed in recent years as a result of the many physiological aspects of cellular metabolism and differentiation in which they are involved. As repeatedly mentioned in previous chapters, polyamine chemical structure is simple; they are aliphatic organic compounds containing amino groups. Ornithine decarboxylase (Odc) or/and arginine decarboxylase (Adc) in plants, spermidine synthase (Spds), *S*-adenosylmethionine (SAM) decarboxylase (Samdc), and spermine synthase (Spms) are the enzymes implied in the biosynthesis of polyamines. The simplest one is putrescine, a compound that originated from the decarboxylation of L-ornithine or arginine in the case of plants by Odc or Adc, respectively. An aminopropyl group is added to putrescine by Spds to form spermidine. Decarboxylated SAM (dcSAM) is the donor of aminopropyl groups, and it is generated from SAM by the action of Samdc. The addition of another aminopropyl group to spermine then leads to the synthesis of spermine, a reaction catalyzed by Spm (Tabor and Tabor 1984). Polyamine back-conversion consists of the transformation of spermine back to spermidine and of spermidine back to putrescine. First, spermidine/spermine-N^1-acetyl transferase (Ssat) transfers an acetyl group from acetyl coenzyme A to the N^1 position of spermine or spermidine. Subsequently, polyamine oxidase (Pao) catalyzes the oxidative deamination of acetylated polyamines into spermidine and putrescine, respectively (Pegg and McCann 1982; Casero and Pegg 1993). For a more detailed description of all these processes, see Chapter 3.

The most intriguing question about polyamines is how they exert their effects, since their mechanisms of action are not yet well understood. However, the comprehension of this fundamental question is related with the physiological roles in which polyamines are involved. In this chapter, some of the most important described functions of polyamines are reviewed.

The ubiquitous and cationic nature of polyamines allows them to interact with negatively charged molecules such as nucleic acids, phospholipids, and some proteins.

But besides these general and unspecific interactions, the fact that they are involved in many aspects of cellular processes, such as cell proliferation, DNA replication, protein synthesis, RNA transcription, and cell differentiation reveals that specific interactions of polyamines with selected molecules must occur. Independently of the large information accumulated regarding the roles of polyamines in cellular processes, it still remains unclear whether the different polyamines have distinct roles in cellular physiology; for this reason, in this chapter, some of the most important described functions of polyamines are reviewed. Evidences indicate that polyamines exert their effects at different levels to alter gene transcription, DNA–protein interaction, protein–protein interaction, and even DNA methylation. Since methylation of specific cytosine residues in eukaryotes alters the binding of regulatory or transcriptional factors (Cedar and Bergman 2009); it is possible to suggest that polyamines affect simultaneously the three phenomena mentioned previously.

On the other hand, with the gradual discovery of polyamine roles in stress tolerance, senescence, differentiation, regulation of plant development, as well as adaptation of different stresses, polyamines have acquired potential at application level, as described in previous chapters, even though it is not known the precise mechanisms by which polyamines have those effects (Kuznetsov and Shevyakova 2007; Alcazar et al. 2010). As some example, the knowledge generated around their physiological roles allows one to propose manipulation of polyamine metabolism as an approach to extend the postharvest life of several fruits such as mango, plum, zucchini, and grapes as well as the increment in production yield of others, such as pistachio, as we will see in this chapter. Exogenous applications of some chemicals, such as nitric oxide, confer salt tolerance in plants such as cucumber seedlings; this effect was correlated with the modification of polyamine levels (Fan, Du, and Guo 2013). Transgenic plants overproducing alkaloids by means of modifications of polyamine metabolism have been also suggested (Bhattacharya and Rajam 2007; Hussain et al. 2011).

As we will see in this chapter, each one of the polyamines seems to have independent roles, and their effects seem even to be opposite under specific conditions; however, examining their effects has been fruitful in the field of postharvest conservation, improvement of plant cultures, and most important, in the field of human diseases and lifespan extension in mammalian models.

6.2 POLYAMINES ARE ESSENTIAL

The essential requirement of polyamines has been thoroughly demonstrated; inhibition of polyamine biosynthesis or mutants affected at any step of polyamine biosynthesis present auxotrophic requirements of polyamines (Smith, Barker, and Jung 1990; Walters 1995). Even the growth of a mutant affected in polyamine retro-conversion (*pao*), which does not require polyamines, is decreased in comparison with the wild-type strain, as it is shown in Figure 6.1. This is the reason why polyamines have been selected as a target to control growth of pathogens or

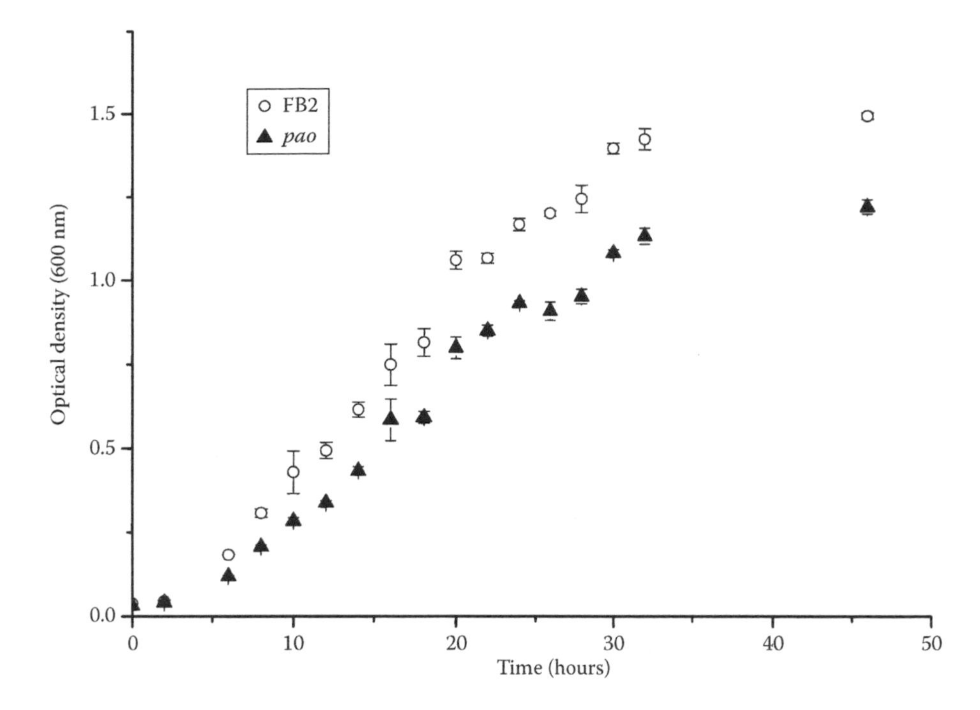

Figure 6.1 Growth rate of a polyamine oxidase mutant (*pao*) of *Ustilago maydis* compared with the wild-type strain (FB2). Cells were grown in minimal liquid medium in triplicate. All points are means ± SE.

even to control tumor growth, as well as manipulation of the cell cycle progression (Rajam and Galston 1985; Devens et al. 2000; Scorcioni et al. 2001; Zaletok et al. 2004).

6.3 ROLE OF POLYAMINES ON CELL GROWTH, TRANSCRIPTION, AND TRANSLATION IN FUNGI AND OTHER ORGANISMS

It is well known that cellular protein synthesis is carried out by ribosomes, complexes formed by two asymmetric macromolecular components made of RNA and proteins, named small and large subunits. In *Escherichia coli*, the small subunit 30S (so-called because it sediments at 30S) is made of 22 components: ribosomal 16S RNA and 21 proteins (S1–S21). The large subunit 50S is made of 34 proteins (L1–L34) and ribosomal 23S and 5S RNA. Biosynthesis and the correct structural organization of ribosomes are of vital importance for their function (Culver 2003; Siibak and Remme 2010), and it has been demonstrated that polyamines are required for the correct biosynthesis and assembly of the 30S particles (Echandi and Algranati 1975; Goldemberg and Algranati 1977; Kashiwagi et al. 1990b).

The transcription of important genes known to be responsible for the control of the cell cycle and proliferation, such as *C-MYC* and *C-JUN* nuclear proto-oncogenes, is also under the control of polyamines (Patel and Wang 1997). In fact, plenty of reports have provided evidence that the polyamine contents of a cell can affect gene expression (Celano, Baylin, and Casero 1989; Celano et al. 1989; Wang et al. 1993, 1999; Wang and Johnson 1994). Recently, a mechanism by which polyamines can alter transcription has been proposed. Through an exhaustive analysis of the promoter region of *SSAT* gene, encoding one of the enzymes involved in polyamine catabolism, the identification of a sequence called polyamine response element (PRE) was possible (Wang et al. 1998). This motif consists of a nine-base pair sequence, 5′-TATGACTAA-3′, is present in the vicinity of promoter of *SSAT*, and appears to modulate *SSAT* transcription. Using the sequence 5′-TATGACTAA-3′ as a probe in an expression library, it was possible to identify a nuclear factor (Nrf-2) that binds constitutively to the PRE sequence (Wang et al. 1998). Nrf-2 had previously been identified as a transcription factor involved in the induction of the transcription of antioxidant enzymes such as γ-glutamylcysteine (Itoh et al. 1997; Jeyapaul and Jaiswal 2000). Later on, by means of a yeast two-hybrid system, a factor designated polyamine modulated factor-1 (PMF1) was found to bind to the leucine zipper region of Nrf-2, resulting in an increase in the transcription of *SSAT* (Wang et al. 1999). Through the use of the yeast two-hybrid strategy, it was found that PMF1 was able to bind to a human homologue of the *Arabidopsis* COP9 signalosome subunit 7a (CSN7) (Wang et al. 2002). The authors suggested that PMF1 might be responsible for the regulation of the transcription of other genes in combination with either Nrf-2 or other potential partners such as Csn 7 protein (Wang et al. 1999, 2001a, 2002). Additionally, it has been reported that there is a competition between Csn 7 and Pmf-1 for binding to Nrf-2. This competition for Nrf-2 binding and interaction with each other was implicated in the regulation of *SSAT* transcription (Wang et al. 2002; Casero et al. 2003).

Csn (Cop9 signalosome complex) is a multiprotein complex of about 500 kDa, containing several polypeptidic subunits, whose function is essentially related with the regulation of protein turnover (von Arnim 2003). In fungi, the existence of the homologous multiprotein complex Cop9 has been described (Braus, Irniger, and Bayram 2010). Csn is essential for the maturation of sexual fruit bodies, as well as for the proper development of hyphae and cell size in *Aspergillus nidulans* (Busch et al. 2003). Csn contains at least two associated enzyme activities, a protein kinase and a deneddylation activity (the latter one consists of the removal of an ubiquitin-like protein from a component of cullin-RING ligases) (Schwechheimer and Deng 2001), which contributes to the regulation of targeted protein degradation. Csn is important for many other functions, and its absence affects transcription of numerous oxidoreductases, and it also leads to hypersensitivity to oxidative stress. Interestingly, Csn is required to activate glucanases and other cell wall recycling enzymes during fungal development (Nahlik et al. 2010). In mice, a mutant unable to produce a Cop9 signalosome subunit Csn8 exhibited a shortened G1 phase duration and showed affected expression of G1 regulators (Liu et al. 2013a). Interestingly, many of the functions in which Csn mutants are affected are similar to those physiological functions affected in polyamine mutants (Paulus, Kiyono, and Davis 1982; Balasundaram, Tabor, and Tabor 1993; Chattopadhyay, Tabor, and Tabor 2002). Thus, it is possible that polyamines could participate in the regulation of a large number of genetic networks that ultimately affect the development of the entire fungal organism, or else that polyamines may be essential to the cross-connections between the different regulatory circuits.

It must be indicated that the PRE sequence 5′-TATGACTAA-3′ has been sought in promoters of different up-regulated genes in tomato fruits overexpressing spermidine and spermine; however, no identical sequences could be found (Srivastava et al. 2007).

In the same token, posttranscriptional regulation of polyamines through the genes encoding spermidine/spermine N^1-acetyltransferase 1 (Sat1) and spermine oxidase (Smox) by micro-RNAs has been reported in prefrontal cortex brain tissue samples (Lopez et al. 2014). These results are suggestive that a different level of regulation by polyamines may depend on gene silencing.

It has been suggested that polyamines regulate translation both at the initiation and elongation steps by means of providing the precursor of the translation factor eukaryotic initiation factor A, a highly conserved and essential protein that also regulates translation elongation in yeast and in mammalian cells through the regulation of translation initiation by means of the modulation of phosphorylation of eIF2α and 4E-BP1, two key regulators of this process (Gregio et al. 2009; Patel et al. 2009; Saini et al. 2009; Landau et al. 2010).

Another effect that may be important for polyamine effects on translation is the observation that they are capable of interfering with protein–protein interactions, since they might efficiently bind to negatively charged motifs at protein interfaces, as was shown with the bovine mitochondrial Cyp11A1 electron transfer system, a three-component complex (Berwanger et al. 2010).

It is well known that phosphorylation of proteins is an essential mechanism in signaling networks that regulate cell function and fate in fundamental cellular processes (Krebs 1994; Hunter 2000; Litchfield 2003). Recent studies suggested that polyamines are able to activate and to regulate the expression and activation of mitogen-activated protein kinases such as p42/p44 in mammalian cells (Bachrach, Wang, and Tabib 2001; Bauer, Buga, and Ignarro 2001). *In vivo*, Odc overexpression leads to the translocation of casein kinase 2 (Ck2) from the cytoplasm to the nucleus, as well as to an increase in its enzyme activity and levels (Shore, Soler, and Gilmour 1997). Odc overproduction also led to an increase in Mapk and tyrosine kinase activities (Auvinen et al. 1992; Flamigni et al. 1999). *In vitro*, polyamines have been shown to stimulate the binding of Ck2 to DNA and the phosphorylation of some Ck2 substrates (Filhol, Cochet, and Chambaz 1990; Gundogus-Ozcanli, Sayilir, and Criss 1999). The evidences showed that spermine interacted directly with the Ck2 β subunit, since stimulation by spermine was lost when glutamine residues 60, 61, and 63 of the β subunit were replaced by alanine residues in COS7 cells. Moreover, a functional domain to which spermine binds made by Asp51-Pro110 was proposed (Leroy et al. 1997).

Other examples of protein–DNA interaction modulated by polyamines have been described. Thus, polyamines increased the binding of estrogen receptor to the estrogen response elements, as well as binding of the nuclear factor kappa B (NF-κB) to NF-κB response element in breast cancer cells (Panagiotidis et al. 1995; Thomas et al. 1995; Shah et al. 1999b; Kuramoto et al. 2003). These results suggested that polyamines might modulate interactions that affect gene expression at the transcriptional level. What unfortunately has not been demonstrated yet is how polyamines or their analogs directly affect the interaction between these elements and if this really affects the transcription of genes involved in differentiation, growth, and proliferation.

6.4 REGULATION OF THE EXPRESSION OF SPECIFIC GENES BY POLYAMINES IN FUNGI AND OTHER EUKARYOTES

In the last few years, global transcriptome analyses have revealed important information about networks related with polyamines. In fact, the multiple effects induced by polyamines are by themselves an evidence of their importance.

DNA microarrays have been employed to unravel the pathways affected by polyamine metabolism, and transcriptome analyses are evidence that polyamines can be modulating the regulation of gene expression (Wang and Johnson 1994; Patel and Wang 1997; Wang et al. 1999). Polyamines have an important effect on global gene expression in fungi, as was shown in *Saccharomyces cerevisiae* double mutants devoid of Spds and Smox, where 10% of the genome responded directly or indirectly to the addition of spermidine or spermine. It was observed that transcription factors related with the metabolism of amino acids and inositol, as well as genes involved in zinc regulation and salt tolerance, were differentially expressed (Chattopadhyay et al. 2009). Most important, this study showed that the cell responds in different ways, depending on the polyamine added to the auxotroph, since the data obtained by addition of spermidine were different from those obtained by spermine addition, showing that spermidine had a major role in the transcriptional expression. The authors highlighted the necessity to compare different mutants to eliminate indirect effects, since the addition of polyamines might affect the regulation by repressing biosynthetic genes, as well as polyamine precursors, such as methionine or arginine (Chattopadhyay et al. 2009). Nevertheless, considering that polyamines are essential for cell viability, it would be expected that genes involved in normal metabolism were differentially expressed; but if we focus on the conditions of the analysis to specific physiological process (stress response, pathogenesis, dimorphism, any differentiation step, etc.), it may be possible to determine the mechanism by which polyamines regulate the expression of specific genes.

Another example of the useful technique of microarray analysis for the comprehension of the mechanism of action of polyamines is *Ustilago maydis*. Transcriptome analysis of an *odc* mutant supplemented with low or high concentrations of spermidine revealed that 2959 genes (about 42%) of the genome were differentially expressed, with 1428 genes being overexpressed and 1532 genes downexpressed (Pérez-Rodríguez and J. Ruiz-Herrera, unpublished results). Besides the genes encoding nonclassified proteins, the categories represented by the largest numbers of genes were those encoding proteins binding to cofactors or proteins, proteins involved in metabolism, and proteins involved in cellular transport. Interestingly, a large number of the up-regulated genes were involved in amino acid metabolism and the synthesis of ribosomal proteins.

In the same way, using the right conditions, it was possible to observe some similarities between the expression of genes involved in protein synthesis, ribosomal assembly, eukaryotic initiation factors (eIF family proteins), as well as transport-related genes (Igarashi and Kashiwagi 2000; Srivastava et al. 2007; Chattopadhyay et al. 2009).

Tomato fruit overproducers of spermidine and spermine were used to analyze the global gene expression profile in an array composed of 1066 unique fruit cDNAs. The results obtained showed that 25% of the genes were differentially expressed in the presence of polyamines, with several pathways being represented, among them genes related with stress and defense response, protein biosynthesis and degradation, and signal transduction (Srivastava et al. 2007).

Arabidopsis transgenic plants overexpressing *SAMDC1* and *SPMS* genes showed elevated levels of spermine, and comparison of transcriptome profiles under different experimental conditions with wild-type Col-0 plants displayed 234 up-regulated and 333 down-regulated common genes. Interestingly, among them were genes involved in abiotic stress responses and in the biosynthesis of jasmonic acid, as well as genes involved in transduction pathways. Plants overexpressing *ADC2* (encoding Adc) that accumulated putrescine were also compared with the wild-type Col-0 and with mutants with elevated levels of putrescine and spermine. In these analyses, the presence of 150 genes regulated in common was observed. Most of them were genes of stress response, as well as genes related with calcium signaling (Marco et al. 2011). These results clearly provide evidence that as a consequence of a disturbance in polyamine homeostasis, there occurred important modifications of transcriptional expression of a large number of genes.

6.5 REGULATION BY POLYAMINES OF GENE EXPRESSION BY MODIFICATION OF DNA METHYLATION PATTERNS

DNA methylation and histone posttranslational modifications lead to the recruitment of protein complexes that regulate transcription (Berger 2007). A hypothesis about the mechanism of action of polyamines is connected with their relationship with DNA methylation. DNA methyltransferases and polyamines share a common substrate, SAM. In fungi, as well as in other eukaryotes, the modification of DNA by methylation affects gene activity and differentiation (Antequera et al. 1984; Jupe, Magill, and Magill 1986; Magill and Magill 1989; Zhu and Henney 1990; Cano-Canchola et al. 1992). It has been suggested that polyamine levels affect gene expression through DNA methylation (Cano-Canchola et al. 1992). Using *Mucor rouxii* as a model system, it was observed that the addition of 1,4-diaminobutanone (DAB), an inhibitor of Odc, inhibited spore germination. In its presence, spores became spherical and increased in size but were unable to form germination tubes (Calvo-Mendez,

Martinez-Pacheco, and Ruiz-Herrera 1987). Interestingly, when the cytosine methylase inhibitor 5-azacytidine was added to spores treated with DAB, germination took place (Cano, Herrera-Estrella, and Ruiz-Herrera 1988). In this context, *in vitro*, polyamines were found to be able to inhibit cytosine-DNA methyltransferases probably by affecting DNA enzyme interaction (Ruiz-Herrera, Ruiz-Medrano, and Domínguez 1995). That this effect was specific, and not merely due to binding of polyamines to DNA, thus inhibiting the access of the enzymes to the substrate, was demonstrated by the observation that the activity of the corresponding restriction enzymes that recognize the same sequences was not affected by polyamines. Moreover, adenine DNA methylases were not inhibited by polyamines. Further studies showed that in the pathogenic fungus *Pyrenophora avenae*, a synthetic putrescine analog (E)-1,4-diaminobut-2-ene reduced DNA methylation (Walters 1997). Additionally, a study involving three dimorphic fungi, *M. rouxii*, *Yarrowia lipolytica*, and *U. maydis*, showed a correlation between DNA methylation, fungal dimorphic transition, and polyamines (Reyna-Lopez and Ruiz-Herrera 2004).

The relative levels of SAM and dcSAM during polyamine synthesis may regulate methylation, since DNA methylases and Samdc share SAM as a common substrate (Ruiz-Herrera 1994; Fraga, Rodriguez, and Canal 2002). For this reason, SAM accumulation is related to an increased methylation of low-molecular-weight compounds, nucleic acids, and proteins (for a review on this topic, see Chiang et al. 1996; Mato et al. 1997; Lu 2000; Bottiglieri 2002; Lieber and Packer 2002; Fontecave, Atta, and Mulliez 2004; Loenen 2006).

It is opportune to recall that Samdc is the enzyme in charge of the decarboxylation of SAM and generation of dcSAM, the aminopropyl donor necessary for the formation of spermidine and spermine. Once SAM is decarboxylated, it is inactive as methyltransferase substrate and therefore leads to a reduction in methylation reactions (Loenen 2006). Accordingly, an organism defective in the generation of Samdc would be expected to accumulate high levels of SAM, the substrate of the enzymes that would be available for an increase in methylation reactions. When *U. maydis samdc* mutants were compared with *U. maydis spds* and *odc* mutants, it was observed that *samdc* mutants accumulated high concentrations of SAM, compared with the wild-type strain, whereas *spds* mutants accumulated SAM levels corresponding only to half the concentration of *samdc* mutants. Nevertheless, dcSAM was found to be accumulated in the *spds* mutants (12-fold increase comparatively with the FB2 wild-type strain, while, as expected, it was absent in *samdc* mutants; Valdes-Santiago et al. 2012b). Interestingly the behavior of both types of mutants was different; *spds* mutants possessed a low level of virulence in maize and lost its dimorphic capacity, whereas *samdc* mutants were completely avirulent, dimorphic, and more sensitive to LiCl addition (Valdes-Santiago et al. 2012b). These differences may be related to their different proportions of SAM and dcSAM. It may be recalled that it has been established that Samdc is not only critical for polyamine biosynthesis but also plays a key role in determining the disposition of the cellular SAM pools (Pegg et al. 1998) and, accordingly, in methylation levels.

Additionally, a negative relationship between the levels of dcSAM and the state of DNA methylation and a positive relationship to cell differentiation have been established in other systems (Frostesjo et al. 1997). Examples of the effects of dcSAM are the reports of Duranton et al. (1998), who described that treatment of a CaCo-2 cell line (human colon carcinoma cells) with an inhibitor of *Samdc* gave rise to an increase in global DNA methylation and the expression of a differentiation marker. Additionally, it was reported that depletion of polyamine biosynthesis in F9 teratocarcinoma stem cells produced an increase in the level of dcSAM, leading to induction of differentiation. This effect was counteracted by the specific inhibition of *Samdc* (Frostesjo et al. 1997). In connection with the effect of SAM, it was reported that if its synthesizing pathway was blocked by 3-deaza-adenosine, hypomethylation and differentiation of muscle were promoted (Scarpa et al. 1996). With these data, Fuso et al. (2001) suggested the possibility of silencing genes regulated by DNA methylation through the administration of exogenous SAM.

Previously, a correlation between protein methylation, polyamines, and morphogenesis in *Mucor racemosus* was found (Garcia et al. 1980). It is also important to cite that in models such as a hamster oral cancer cell line, the ectopic expression of Odc antizyme led to a reduction in the Odc activity, and an increase in demethylation of 5′methyl cytosines was observed (Tsuji et al. 2001; Yamamoto et al. 2010).

6.6 POLYAMINES AND THE STRUCTURE OF CELL MEMBRANES

Interactions of polyamines with the negatively charged phospholipids of membranes are given by the positive charge of polyamines at physiological pH. But interestingly, even sterols, which are present in the membranes of eukaryotic organisms including fungi (Alvarez, Douglas, and Konopka 2007; Kristan and Rizner 2012; Vida et al. 2012) and have no negative charge, appear to be involved in the interaction with polyamines. Thus, the isolation of a sterol conjugate with spermine has been reported (Moore et al. 1993). Polyamines at physiological concentrations reduce membrane fluidity, as demonstrated by the observation that putrescine and spermidine destabilized preformed lipid membranes (Zheliaskova, Naydenova, and Petrov 2000).

It is possible that polyamines, being aliphatic and strongly positively charged molecules, are capable to bind to ligands and hence, in some way, facilitate their binding to membrane receptors, as was shown with insulin receptor. In this case, polyamines enhanced the binding of the insulin receptor to the plasma membrane in Langerhans islet β cells (Sandler et al. 1989). This effect may occur through the interaction of polyamines with the phospholipids of the plasma membrane (Meers et al. 1986; Sandler et al. 1989).

Interestingly, and independently of the data described previously, the physiological significance of the interaction of polyamines with components of the cell membranes remains to be clearly established. As a feasible hypothesis, their direct interaction with G proteins as a mechanism for the action of polyamines has been proposed, regulating in this way receptor-dependent processes (Bueb, Mousli, and Landry 1991). In this sense, it is important to indicate that polyamine depletion in intestinal epithelial cells does not alter the activation of Rho GTPases by guanine nucleotide exchange factors (GEFs) (the family of proteins responsible for catalyzing nucleotide exchange in the Rho GTPases); as a result, it may be interpreted as an evidence that polyamines may act upstream of GEF activation (Ray et al. 2011). In the study of membranes, the process of rectification refers to their property to show changes of conductance according to the voltage applied. In this sense, it has been observed that cytoplasmic spermine and spermidine are able to block membrane channels of *Xenopus oocytes* cells, resulting in a strong inward or anomalous rectification (Lopatin, Makhina, and Nichols 1995). It is known that, in general, inward rectification results from pore blocking, thus indicating that this is precisely the effect caused by intracellular polyamines (Hibino et al. 2010). Some evidences suggested that spermine has a major role in the rectification of potassium channels, with the biological implication of the modulation of receptors such as the *N*-methyl-D-aspartate receptor (Johnson 1996; Williams 1997a,b, 2009; Kurata, Marton, and Nichols 2006). In the same way, polyamine uptake has been associated with increases in the levels of cyclic adenosine monophosphate, suggesting that this second messenger system might be involved in the phenomenon (Khan, Quemener, and Moulinoux 1991).

In plant cells, polyamines affect ion channels and pumps controlling cationic transport across membranes. The physiological significance of this phenomenon is related with ion homeostasis, signaling, and plant response to stress and it has been extensively reviewed (Pottosin and Shabala 2014). In *Zea mays* seedlings, it has been reported that putrescine induces an elevation in the activity of phospholipase D during the later stage of drought stress response, which produces membrane damage. This result was interpreted as an outcome of the imposed injury (An et al. 2012).

6.7 THE ROLE OF POLYAMINES IN STRESS RESPONSE

As indicated previously, in fungi, polyamines have been cited to be involved with stress protection. Thus, increasing reports have been published demonstrating that an increase in the levels of polyamine concentrations occurs in response to diverse types of stress. Accordingly, it has been shown that a large number of genes are differentially regulated when fungi respond to environmental stress and that many of these genes are conserved among different species, such as those involved in carbohydrate metabolism, protein metabolism, defense against reactive oxygen species (ROS), intracellular signaling etc. (Giaever et al. 2002; Gasch 2007). These responses have been shown to be affected by polyamines, as observed in fungal mutants affected in polyamine metabolism. In *S. cerevisiae*, evidence for this effect was obtained when it was demonstrated that polyamines protected the cell and its components from oxidative insults (Balasundaram, Tabor, and Tabor 1993; Chattopadhyay, Tabor, and Tabor 2006a; Watanabe, Watanabe, and Kondo 2012). These authors observed that an Samdc (*samdc*) mutant was more sensitive than the wild type to oxidative damage, suggesting that one possible function of polyamines was to protect cell components from oxidative stress (Balasundaram, Tabor, and Tabor 1993). In general, the ability to deal with stress is diminished in mutants affected in any of the genes encoding enzymes related with polyamine metabolism. In this regard, *U. maydis* is a well-studied system since it has been demonstrated to be an important model organism; besides, in contrast to *S. cerevisiae*, a common model of study, *U. maydis* contains only putrescine and spermidine and, as already pointed out, lacks spermine (Valdes-Santiago, Cervantes-Chavez, and Ruiz-Herrera 2009). In a comparative study, single and double mutants of *U. maydis* affected in genes involved in polyamine metabolism (*odc/spds*, *odc/samdc*, *odc/pao*, *odc*, *samdc*, and *spds*) were grown on solid media containing 1.5 M KCl or 1 M NaCl. The results obtained showed that all of them were affected in their dealing to stress conditions since the ability to grow was decreased compared with the wild-type strain (Valdes-Santiago and Ruiz-Herrera 2014). Furthermore, when Pao null mutants were compared with the double mutant *odc/pao* unable to produce putrescine by either direct or retro conversion, it was observed that only the double mutant was susceptible to media containing a high concentration of NaCl, LiCl, or sorbitol. Another comparative study performed with spermidine auxotrophic mutants revealed some specificity of the polyamines with regard to stress resistance to high concentrations of positive ions. Growth of both *samdc* and *spd* mutants was inhibited under high levels of NaCl, but *spd* mutants were insensitive to LiCl concentrations, which inhibited the growth of the wild-type strain (Valdes-Santiago et al. 2012b).

Interestingly, under oxidative stress induced by H_2O_2, a *pao* mutant was sensitive, but an *odc/pao* mutant showed resistance. In light of this result, we suggested that *in vitro*, spermidine was more effective than putrescine in affording protection to oxidative stress (Valdes-Santiago, Guzman-de-Pena, and Ruiz-Herrera 2010). Also, spermine seems to be involved in oxidative protection since the addition of spermine reverted the lipoperoxidation induced by H_2O_2 in *Trypanosoma cruzi*, and spermine reverted the apoptosis induced by oxidative stress in *spds/pao S. cerevisiae* mutants (Chattopadhyay, Tabor, and Tabor 2006a; Hernández, Sánchez, and Schwarcz de Tarlosvsky 2006).

As indicated in a previous chapter, plant subjected to different stress conditions respond usually, although not always, with an increase in polyamine levels, which has been interpreted as a reaction of protection (Alcazar et al. 2006, 2010; Gill and Tuteja 2010; Gupta, Dey, and Gupta 2013). This effect is due to the fact that genes related with polyamine metabolism are transcriptionally induced during different kinds of stress (Mutlu and Bozcuk 2007; Liu, Inoue, and Moriguchi 2008). Something similar takes place in cyanobacteria, where an elevation in spermine and spermidine levels is observed under salt stress and osmotic stress, respectively. In addition, an increase in arginine decarboxylase level and in the uptake of putrescine and spermidine took place (Jantaro et al. 2003; Incharoensakdi et al. 2010). Nevertheless, despite these considerations, there are few specific molecular functions of polyamines that have been reported related with stress. One of the few examples where a positive correlation was observed refers to the fact that spermine is considered the major natural intracellular compound capable of protecting DNA from free radical attack. Accordingly, it was documented that spermine was capable of reacting directly with ROS, thus reducing radical attack (Ha et al. 1998a,b). Nevertheless, it may be recalled that a large number of fungi do not contain spermine and are able to respond to stress through alterations in the metabolism of polyamines (see Valdes-Santiago and Ruiz-Herrera 2014).

The protective effect of polyamines to stress also occurs in bacteria. Evidences demonstrated that the addition of putrescine protected *E. coli*, both *samdc/odc* double as well as single mutants deficient only in the *SAMDC* gene, against oxidative conditions (Chattopadhyay, Tabor, and Tabor 2003b). In agreement to this result, exogenous putrescine addition alleviated ROS damage induced by the antimicrobial peptide polymyxin in *Burkholderia cenocepacia* (El-Halfawy and Valvano 2014). Also, an Spds mutant of *Pseudomonas aeruginosa* showed an increment in sensitivity to H_2O_2, as compared with the wild-type strain. Interestingly, this mutant did not present spermidine and putrescine at the outer membrane, and the authors suggested that the absence of these polyamines led to the susceptible phenotype under treatment with polymyxin B. In addition, the expression of Spds and Samdc was induced by the addition of membrane damaging substances (Johnson et al. 2012).

In vitro, polyamines have been reported to radioprotect eukaryotic DNA (Chiu and Oleinick 1998; Douki, Bretonniere, and Cadet 2000), and this effect has been attributed to the induction of DNA (or chromatin) compaction and aggregation. Moreover, polyamine depletion induced cell radiosensitivity (Williams, Casero, and Dillehay 1994; Spotheim-Maurizot et al. 1995; Newton et al. 1996). Deproteinized human cell nuclear DNA is preserved by spermine under radiation conditions, but interestingly, putrescine showed less protection than spermine did, which was found to bind to the nucleus, indicating that spermine might be the polyamine conferring the resistance to radiation damage (Walters et al. 1999). In agreement with these data, spermine and spermidine showed more radioprotective effect than did putrescine of the pBR322 plasmid DNA; however, radioprotection was explained in terms of scavenging of radicals and reduction of the accessibility of the sites susceptible to attack (Spotheim-Maurizot et al. 1995).

The results presented in this section showed that each polyamine behaves differently in their protective characteristics to different types of stress. However, most of the studies on polyamines have not taken into consideration that each one of the polyamines might posses a specific role at the biochemical level, moreover when it is known that each polyamine is different in charge and structure (van Dam, Korolev, and Nordenskiold 2002).

6.8 SELECTED ASPECTS OF THE ROLE OF POLYAMINES IN SOME SPECIFIC CELLULAR FUNCTIONS

One of the most studied functions of polyamines is their requirement for cell growth. However, the molecular mechanisms behind this effect are largely unknown. In this regard, it has been hypothesized that one of the most important biochemical reactions in which spermidine is involved, at least in the case of eukaryotic organisms, is their role as precursor of hypusine, an uncommon amino acid that binds to the eukaryotic translation initiation factor 5A (eIF5A) (Park, Wolff, and Folk 1993a). Deoxyhypusine synthase (Dhs; EC 2.5.1.46) transfers a 4-aminobutyl moiety from spermidine to the ε-amino group of a specific lysine residue in an eIF5A precursor protein in the presence of NAD^+. To generate the biologically active eIF5A, hydroxylation of the deoxyhypusine residue by deoxyhypusine hydrolase (Dohh) is necessary (EC 1.14.99.29) (Park and Wolff 1988; Park et al. 2006). *S. cerevisiae* haploid mutants defective in the gene encoding Dhs produce a lethal phenotype, revealing its essentiality for the cell growth of yeast (Wolff, Park, and Folk 1990; Sasaki, Abid, and Miyazaki 1996). Interestingly, an *S. cerevisiae* conditional mutant that expressed an unstable eIF5A presented no significant difference in the rate of protein synthesis compared with wild type, which indicated that global protein synthesis is not the main role of eIF5A. However, in the same study, eIF5A-depleted cells displayed a defect in cell division, impeding cell proliferation (Kang and Hershey 1994). Additionally, it was described that *S. cerevisiae* polyamine auxotrophic mutants deficient in the synthesis of Odc, Spds, or Samdc, when growing under low concentration of spermidine used up to 54% of this polyamine for hypusine modification of eIF5A (Chattopadhyay, Park, and Tabor 2008). It is important to mention that acetylation of eIF5A at the hypusine residue by Ssat inactivates the protein, regulating in this way its activity without a change at the protein level (Park, Cooper, and Folk 1981; Park, Wolff, and Folk 1993a,b). Among the reported functions of eIF5A is the regulation of mRNA stability, translation elongation, RNA export, and apoptosis (Ruhl et al. 1993; Schrader et al. 2006; Saini et al. 2009; Caraglia et al. 2013). In *Fusarium*, *DHS* gene transcription was induced in infected wheat spikes under nutritional stress, but not in wheat spikes in the course of normal vegetative growth, preventing the *Fusarium* head disease (Woriedh et al. 2011). Regarding Dohh, it seems not to be essential in *S. cerevisiae* since defective mutants presented only diminished growth (Park et al. 2006). However, in *Schizosaccharomyces pombe*, a mutation in the Dohh homologue caused severe abnormal mitochondrial distribution and morphology, as well as misorientation of microtubules after extended incubation at 37°C (Weir and Yaffe 2004).

In relation with this matter, it is interesting to cite our data where mutants unable to synthesize spermidine were able to grow in the presence of spermine (Valdes-Santiago, unpublished results). The possible explanation for this result and its consequences regarding the necessity of hypusine biosynthesis were discussed in Chapter 4.

6.9 NATURAL POLYAMINES AND THEIR POSSIBLE CLINICAL AND BIOTECHNOLOGICAL APPLICATIONS

Although not directly related to polyamine metabolism in fungi, we have considered important to review some empirical applications of polyamines for the sake of completeness and because the results obtained with other eukaryotes might be, at some time, applicable to fungal diseases.

6.9.1 Polyamines and Their Potential Medical Applications

6.9.1.1 Selected Topics of Polyamines in Cancer

In the last few decades, studies on the association of polyamine metabolism and cancer have been an area extensively investigated (Gerner and Meyskens 2004; Casero and Marton 2007; Magnes et al. 2014). Polyamine analogs such as bis(ethyl)norspermine have proved to be efficient in the reduction of tumors and tumor growth on transgenic mice presenting mammary cancer (Shah et al. 1999a,b). Other strategies that were suggested to be used in the control of different tumor cells were based on the observation of a correlation between polyamines, apoptosis, and transglutaminases (Wang, Viar, and Johnson 1994). Transglutaminase activity and polyamines in programmed cell death have been related, since in mammalian cells exposed to spermidine, transglutaminase activity was increased and at the same time apoptosis was also induced. In cell lines IEC-6 and Caco-2, polyamines not only increased transglutaminases activity but also affected cell migration, and interestingly, it was found that this effect was reverted by the addition of polyamine inhibitors as well as inhibitors of transglutaminase activity (McCormack et al. 1994; Facchiano et al. 2001). Polyamines are substrates for transglutaminases, and (γ-glutamyl) putrescine, N^1-(γ-glutamyl) spermidine, and N^8-(γ-glutamyl) spermidine have been identified to be products of this reaction (Williams-Ashman and Canellakis 1979; Folk et al. 1980). In the study of transglutaminases in various cancer cell lines, it was reported that there occurs an elevation of protein–polyamine complexes when transglutaminase activity is stimulated, with induction of apoptosis (Beninati, Abbruzzese, and Cardinali 1993). The ability of transglutaminases to generate protein–polyamine conjugates, which in turn would induce apoptosis in cancer cell lines, has led to the suggestion of the use of transglutaminase activators, such as phytochemicals like all-trans acid (retinoid), epigallocatechin-3-gallate, and curcumin (flavonoids), as antineoplasic agents (Lentini et al. 2009, 2010).

Early detection of cancer has been one of the goals of researchers in the last decades. Several lines of evidence point out that abnormal polyamine metabolism is considered not only as a signal of development of cancer but also that polyamine content in serum, plasma, and/or urine might be a biomarker of cancer (Suh et al. 1997; Teti, Visalli, and McNair 2002). It has been shown that patients at the beginning of cervix cancer development, in comparison with normal subjects, presented no differences in polyamine levels, including putrescine, spermidine, spermine, cadaverine, *N*-acetylputrescine, *N*-acetylcadaverine, N^8-acetylspermidine, N^1-acetylspermidine, N^1-acetylspermine, N^1, N^{12}-diacetylspermine, and 1,3-diacetylpropane (Lee et al. 2003). Nevertheless, an association of the elevation of concentration of the polyamines mentioned previously occurred once cervical cancer was established. Specially, putrescine, spermidine, and spermine were found significantly increased in cancer subjects (Lee et al. 2003). Plasma and urine from hepatic cancer showed similar patterns with an elevation in the level of spermidine, spermine, and *N*-acetylspermidine, whereas

L-ornithine, putrescine, and γ-aminobutyric acid were found in minor concentration, compared with the plasma of healthy subjects (Liu et al. 2013b).

Determination of the levels of polyamines as a biomarker seems to be an attractive possibility in the detection of different kinds of cancer; however, the probability of false-positives is great, considering other physiological states where polyamines present high levels in plasma and/or urine, such as pregnancy and amyloidogenic diseases among others (Russell et al. 1978; Saito 1985; Khuhawar and Qureshi 2001; Moinard, Cynober, and De Bandt 2005; Chowhan and Singh 2013).

Female breast cancer patients present high level of spermidine and spermine, and this elevation was correlated with the grade of apoptosis found in the tissue samples compared with healthy ones (El-Salahy 2002). One possible explanation of how high polyamine concentration could induce apoptosis in different human tumors is given by the action of amine oxidases, enzymes involved in polyamine level regulation, which generate H_2O_2 and aldehydes as reaction products, which in turn are implicated in cell death (Agostinelli et al. 2004). For these reasons, it has been proposed that the use of amine oxidases in combination with natural polyamines could have potential therapeutic applications. This possibility is supported by the observation that a combination of bovine serum amine oxidase (Bsao) and spermine is able to induce a cytotoxic effect given by oxidative stress, produced in turn by the action of H_2O_2 and aldehydes, as well as acrolein (Calcabrini et al. 2002). In this regard, human epidermoid KB and breast cancer MCF-7 cell lines were treated with a combination of Bsao/spermine and docetaxel (an inductor of apoptosis). As a result, the fluorescent microscopy assay showed the appearance of 36% of apoptotic cells. Western assays showed the activation of caspases 9, 3, and 8 and the activation of the mitochondrial apoptotic pathway (Marra et al. 2008; Agostinelli 2012). The development of nanoparticles made with a nucleic acid assembly together with BSAO and spermine has been recently reported (Agostinelli et al. 2010, 2015). This strategy has a promising potential as anticancer therapy.

6.9.1.2 Polyamines and Their Potential Application in Systemic Lupus Erythematosus

As we make progress in our knowledge of the polyamine physiological roles, it is possible not only to use the information obtained for our comprehension of polyamines per se but also to apply it for using polyamines in the treatment and control of diverse diseases, as described in Section 6.9.1.1 (Agostinelli et al. 2004; Gugliucci 2004; Ramani, De Bandt, and Cynober 2014). The physicochemical properties of polyamines have been reviewed in Chapter 2, where it has been stressed that their cationic nature contributes in their interaction with DNA, leading to modification of its conformation. Some diseases, such as systemic lupus erythematosus (SLE), have been related with anomalous changes in DNA conformation, giving rise to left-handed Z-DNA, an imunogenetic form that binds to anti-DNA antibodies or even causes disruption of the chromatin structure (Thomas, Meryhew, and Messner 1990; Brooks 1994). SLE is an autoimmune disorder characterized, as mentioned previously, by the generation of anti-DNA antibodies in the sera of patients (George and Tsokos 2011). Exogenous addition of 1 mM spermine to plasma from SLE patients resulted in a dose-dependent inhibition of the interaction between DNA and anti-DNA antibodies and affected the stability of preformed complexes. Spermine probably blocks phosphate groups, and thus, its interaction could be able to prevent the union with anti-DNA antibodies. Therefore, the use of polyamines or their derivatives has been proposed to avoid the generation of immune complexes, improving in this way the heath of the patients (Wang et al. 2014a).

6.9.1.3 Polyamines and Their Potential Application on Other Affections

6.9.1.3.1 Polyamines and Their Application on Metabolic Syndromes and the Extent of Life Span

Early findings showed that spermine induced oxidation of glucose to carbon dioxide in isolated rat adipocytes, imitating insulin action. This effect is due probably to its interaction with receptors of insulin-responsive tissues at the level of the cell membrane (Lockwood and East 1974). Accordingly, spermine and its analogs have been proposed in the treatment of metabolic syndromes.

In this line, obese mice exposed to intraperitoneal 10 mg/kg spermine lost 24% bodyweight, and this effect was related with increasing glucose metabolism and fatty acid oxidation (Sadasivan et al. 2014). In another study, it was found that the oxidation of glucose and lipids was also stimulated by arginine (the precursor of natural polyamines; see Figure 3.1) supplemented in a diet with atlantic salmon. The isolated liver cells presented high activity of Ssat and differential expression of genes implicated in apoptosis and polyamine biosynthesis (Andersen et al. 2014). The consequences of arginine oral intake in mice included elevation in spermidine concentration in blood and increased production of polyamines by the microbiota (Kibe et al. 2014). In this latter study, the researchers attempted to elevate polyamine concentration with the goal of increasing mice longevity, taking into consideration their role as anti-inflammatory agents, inhibitors of oxidative stress, and promoters of gut bacteria (Lagishetty and Naik 2008; Eisenberg et al. 2009; Matsumoto et al. 2011). The conclusion was that arginine supplementation plus LKM512 (a probiotic bifidobacterium) delivered the best result in preventing inflammation in body and brain and degenerative age-related diseases (Kibe et al. 2014). Knowledge of the mechanism behind the effect of polyamines over the lifespan is of utmost importance; however, the prolongation of lifespan has been observed just in simpler organisms, such as *S. cerevisiae*, and *Caenorhabditis elegans* (Madeo et al. 2010). Exogenous application of spermine to aging yeast induced the inhibition of histone acetyltrasferases, leading to the deacetylation of histone H3, as well as an increase in the transcription of genes related with autophagy (Eisenberg et al. 2009). The role of polyamines in longevity can be related with their ability to induce stress resistance, which has been well documented in plants and fungal models (Alcazar et al. 2010; Valdes-Santiago and Ruiz-Herrera 2014). An important contribution of *S. cerevisiae* and *E. coli* to the comprehension of polyamine transport in the finding of human homologues is worth mentioning (Minois, Carmona-Gutierrez, and Madeo 2011); such information supports the idea of fungi as a valuable model to unravel the mechanism of action of polyamine.

Based on evidences indicating that polyamines are present in high concentration in intestinal mucosal cells, and their contribution to the integrity and maintenance of gut function (Guo et al. 2003; Seiler and Raul 2007), spermine has been considered as auxiliary in oral formulation of poorly absorbed drugs. The combination of 10 mM spermine with 25 mM sodium taurocholate showed a synergistic effect for the gastrointestinal absorption of rebamipide in rats (Miyake et al. 2006). In addition, regarding low oral bioavailability, it has been also reported that conjugates with bile acid-polyamines of the important antiviral 3′-Azido-3′-deoxythymidine (AZT, Zidovudine) increased their ability to cross the phospholipid bilayers (Salunke, Hazra, and Pore 2003; D. Wu et al. 2007).

6.9.1.3.2 Polyamines and Their Potential against Pathogen-Induced Diseases

Bacillus anthracis is the causal agent of the infectious disease anthrax, and the inhalation of its spores in most cases is lethal (Hanna 1998). An important virulence factor in the pathogenesis of *B. anthracis* is its lethal factor (LF). This protein is a 90-kDa zinc-dependent metalloprotease (Pannifer et al. 2001; S.L. Johnson et al. 2006). Interestingly, between its several small-molecule inhibitors are polyamines, since their interaction with the LF anionic sites changes the protein conformation and thus inhibits its activity. Accordingly, it was found that unlike spermidine, spermine is able to inhibit LF activity in a concentration depending manner (Goldman et al. 2006).

Use of polyamine derivates has been proposed for the treatment of other illness. Branched polyamines such as polyamidoamine and polypropyleneimine dendrimers also have resulted in potential tools for the treatment of fatal diseases caused by prion proteins, which are distinguished by their infectious potential (Supattapone et al. 2001; MacKintosh, Tabrizi, and Collinge 2003; Jackson 2014). Finally, it must be cited that a spermidine conjugate, tetracyclic 1,2,4-tri-oxane artemisinin, attached at the spermidine amino centers, showed activity against *Plasmodium falciparum* and promyelocytic leukaemia HL-60 cells (Chadwick et al. 2010).

6.9.2 Polyamine Application as Protective Agent in Plant Development

Enough evidence, discussed in previous chapters, demonstrates that plant differentiation, growth, embryogenesis, flowering, stress response, and even apoptosis are phenomena related with polyamines (Kakkar et al. 2000; Alcazar et al. 2010; Do et al. 2014). This fact has led to the consideration that manipulation of polyamine levels might allow the control of biological processes or the responses to different stimuli in plants. For example, dormancy is a physiological state of provisionary suspension of any activity related with meristem growth in plants (Lang et al. 1987), which depends on many signals such as temperature and light (Chao et al. 2007).

Interestingly, a break in dormancy can be induced by the addition of inorganic compounds such as Dormex (hydrogen cyanamide), KNO_3 and mineral oil in buds of "Anna" apple trees. This change in the physiological state was correlated with modifications in polyamine levels during treatments. Whereas higher levels of spermine and spermidine were found at the start of the dormancy, lower levels of these polyamines were found during dormancy release. In this study, Dormex application is proposed as an effective option to increase the productivity of Anna apple trees (Seif El-Yazal and Rady 2012). Another example was reported in grape (*Vitis* species), where ripening induced differential expression of genes involved in both polyamine synthesis and metabolism (Agudelo-Romero et al. 2013).

6.9.2.1 Polyamines and Resistance to Cold Stress Conditions

In plants, membrane integrity and lipid peroxidation are used as indicators of stress levels. Under chilling stress, polyamines are able to protect membrane from oxidation, by increasing the activity of enzymes such as catalase, superoxide dismutase, and peroxidases (Zhang et al. 2009; Koushesh saba, Arzani, and Barzegar 2012). The exogenous addition of 1 mM putrescine or 0.5 mM spermidine increased the chilling resistance of cucumber seedlings (*Cucumis sativus*) by preventing the generation of H_2O_2 (Zhang et al. 2009). The same effect was observed in apricot, since treatment of the fruit with 1 mM putrescine or spermidine not only decreased ethylene production but also maintained the integrity of the fruit and induced the activity of enzymes involved in antioxidant response during storage under cold temperature (Koushesh saba, Arzani, and Barzegar 2012). An increase in antioxidant enzymes has been also observed with the addition of spermidine to seedlings of *Panax ginseng* and *C. sativus* under ionic stress, enhancing salt tolerance (Duan et al. 2008; Parvin et al. 2014).

Long periods of cold storage affect fruit quality; however, low-temperature storage is necessary to avoid microbial contamination and spoilage, changes in fatty acid composition, nitrogen metabolites, carbohydrate and ethylene levels, as well as changes in ascorbic acid metabolism among others (Luengwilai and Beckles 2013; Tsaniklidis et al. 2014). It has been reported that polyamine metabolism in chilling-resistant zucchini fruit could be related with putrescine degradation, proline accumulation, and probably, with the GABA (4-aminobutyrate) catabolism. Polyamine levels as well as the expression of the genes involved in their synthesis and catabolism were analyzed during postharvest storage at low temperatures during conservation of two varieties of *Cucurbita pepo*, which had different chilling tolerance, to uncover the contribution of polyamines to the preservation phenomenon of fruits (Palma et al. 2014). Free putrescine levels decreased when fruits were stored at –20°C, whereas this polyamine increased its levels when fruits were kept at 4°C in both varieties, more in the sensitive one. The authors suggested that the increase in putrescine levels after harvest in cold-sensitive varieties was caused by the cold stress response, instead of the postharvest storage itself, as it was supposed, and that this effect was not a mechanism of protection, but a response to cold storage. Degradation of putrescine and proline accumulation could be important factors that contribute to chilling tolerance, as well as GABA catabolism, which could also be involved in the chilling tolerance of zucchini fruit (Palma et al. 2014). In summary, putrescine seemed not to be involved in the protection of fruit during low-temperature storage, since putrescine levels increased when fruits were kept at 4°C, and the opposite happened at 20°C. This result suggested that putrescine accumulation was the result of response to cold stress. In this regard, spermidine and spermine, at both temperatures, decreased their concentration. The expression of *ADC*, *ODC*, and *PAO* genes was induced under cold storage at 4°C, but interestingly, GABA levels decreased, correlated with higher activity of GABA transaminase during low-temperature storage (Palma et al. 2014). The exogenous addition of 1 mM putrescine, spermidine, or spermine resulted in protection of chilling damage in zucchini fruit (*C. pepo*); nevertheless, putrescine-treated fruits presented less weight lost and chilling injury index (Palma et al. 2015).

Other examples showed that dip treatment, specifically of grapes (*Vitis vinifera* cv *Flame seedless*), with 1 mM spermine during 5 minutes maintained fruit quality and extended life storage at low temperature (Harinda Champa et al. 2014). Application of 1 mM putrescine or spermidine to pomegranate (*Punica granatum*) by immersion or pressure infiltration extended the fruit quality under cold storage (Mirdehghan et al. 2007). Pressure infiltration of 0.25 mM spermidine or 1 mM spermine to apples (*Malus domestica*) prevented softening under cold storage and elevated the firmness (Kramer, Wang, and Conway 1991). A detailed study of changes in the metabolism of *C. pepo* given by the treatment with putrescine showed accumulation of glutamic acid and degradation of GABA, as well as the accumulation of high levels of alanine and ATP content (Palma et al. 2015).

6.9.2.2 Polyamines as Agents for Improvement of Shelf Life

In agreement with most of the studies that show putrescine efficacy to protect fruits during cold storage, exogenous applications of spermine have also shown good effects in the preservation of fruits and other attributes, such as shelf life (Valero, Martínez-Romero, and Serrano 2002). Dip treatment of papaya in spermine (3 mM) or pressure infiltration of spermidine and spermine (0.3, 1, or 3 mM) to mango fruit (*Mangifera indica*) prolonged firmness and shelf life (Purwoko et al. 1998). Treatment of mango fruit cv Kensington pride soaked for 6 minutes in 0.01 mM spermine reduced most efficiently fruit softness compared with treatments with putrescine or spermidine at different concentrations (0.01, 0.5, or 1 mM); weight loss was inhibited most efficiently by spermine 0.5 mM (Malik, Tan, and Singh 2006). Increasing of the production yield of pistachio (*Pistacia vera*) was obtained by exogenous application via spray of spermine (1 mM) and, to a lesser extent, spermidine (1 mM), while putrescine did not have important effects (Khezri et al. 2010).

In addition, exogenous application of putrescine to different fruits with short storage life results in the improvement of fruit quality during postharvest storage. In plum fruit (*Plunus salicina*), treatment with 2.0 mM putrescine extended the storage life without affecting fruit quality, evaluated by total carotenoids, vitamin C, and antioxidants, and exogenous addition of 1 mM putrescine increased fruit firmness and decreased ethylene synthesis (Pérez-Vicente et al. 2002; Khan et al. 2008). In apricot (*Prunus armeniaca*), immersion in different concentrations of putrescine for 5 minutes increased its preservation and quality, and 4 mM putrescine showed to be the best condition to retard the ripening processes (Davary Nejad et al. 2013). In mango cv. Langra, exogenous application of putrescine 2 mM decreased weight loss, as well as ripening retardation, compared with 1 or 3 mM putrescine (Jawandha et al. 2012). Addition of 1 mM putrescine by immersion to mango cv dashehari for 10 minutes extended storage life from 2 to 4 weeks under refrigeration conditions (Bhart et al. 2014).

At the level of fruit ripening, it was demonstrated that polyamines are able to control it by exogenous addition to trees, in nectarines and peaches (*Prunus persica*). In both cases, a reduction in ethylene levels related with down-regulation of genes involved in ethylene biosynthesis was observed, and additionally, the time span of fruit ripening was increased (Bregoli et al. 2002; Torrigiani et al. 2004). Improvement of seed quality induced by the exogenous addition of polyamines has been reported for hot pepper (*Capsicum annuum*), wheat (*Triticum aestivum*), sunflower (*Helianthus annuus*), and rice (*Oryza sativa*); the concentration of added polyamines as well as the most effective ones presented variances between species (Iqbal and Ashraf 2005; Farooq et al. 2007, 2008; Khan et al. 2012). Treatment of hot pepper with aerated solutions of 25 or 50 mM putrescine, rice with 10 ppm putrescine, or wheat with 2.5 mM putrescine enhanced parameters such as grain yield, shoot growth, number of roots, and synchronization of germination (Iqbal and Ashraf 2005; Farooq et al. 2007, 2008; Khan et al. 2012).

6.9.2.3 Polyamines as Agents against Heavy-Metal-Induced Stress on Plants

The role of polyamines in response to heavy metals has attracted attention in light of the increasing contamination by metals of aquatic environments and as a result of the induction of metabolic changes at biochemical and genetic levels suffered by the organisms exposed to them (Groppa, Tomaro, and Benavides 2007; Amoroso and Abate 2012). In *Potamogeton crispus*, a macrophyte, putrescine accumulation and a decrease in spermidine and spermine concentrations were related with Cadmium (Cd^{2+}) stress, and the same changes in the levels of polyamines was presented in *Nymphoides peltatum* under copper (Cu^{2+}) stress (Wang et al. 2007; Yang et al. 2010). The same metabolic changes in polyamine levels were observed also in *Malus hupehensis* treated with Cd^{2+}. In addition, higher activity of Pao and the subsequent production of H_2O_2 seemed to be related with the cell death observed in roots (Jian et al. 2012). Under Cd^{2+} and Cu^{2+} stress, *M. hupehensis* and *N. peltatum*, respectively, showed an elevation in the superoxide anion $\left(O_2^-\right)$, and exogenous addition of spermidine or spermine counteracted this effect (Wang et al. 2007; Mattoo and Handa 2008; Zhao and Yang 2008). Exogenous application of 1 mM spermine to sunflower leaf discs reverted the effect on lipid peroxidation (Groppa, Tomaro, and Benavides 2001) and, when applied to *Raphanus sativus* under chromium stress, modulated the oxidative damage and improved seedling growth (Choudhary et al. 2012).

It has been observed that wheat (*T. aestivum*) plants irrigated with wastewater contaminated with heavy metals lose weight and experience a decrease in soluble sugars, polysaccharides, and total carbohydrates content. However, when wheat grains were soaked in 0.3 mM spermidine, 0.15 mM spermine, or both during 6 hours and were planted and irrigated with heavy-metal-contaminated wastewater, all those effects were counteracted, and the highest protective effect was obtained with the mixture of spermidine and spermine (Aldesuquy et al. 2014).

6.9.2.4 Polyamines as Inducing Agents for Plant Development

In the study of morphogenesis, plant regeneration has been a fruitful system to study. It has been shown that putrescine and spermidine are implied in the early events leading to somatic embryogenesis, and exogenous addition of spermine induces the growth of maize calli (Pedroso et al. 1997; Ahmadabadi, Ruf, and Bock 2007). In different rice genotypes with different abilities of regeneration, it was possible to revert this process by modification in the levels of polyamines or by inhibition of their synthesis using difluoromethylarginine, an inhibitor of arginine decarboxylase. This is an interesting effect since it would allow modified plant regeneration ability in recalcitrant species (Shoeb et al. 2001). In the development of tissue cultures for plant regeneration, manipulation of cytokinin has been suggested as a mean to induce plant regeneration (Hill and Schaeller 2013); however, polyamines also have significance. Thus, in plants such as *Lasiurus scindicus*, *Sorghum halepense*, and *Urochloa panicoides* regenerating from young inflorescences, polyamine optimum levels have been determined in embryogenic and nonembryogenic cultures, with the goal of eventually modulating the embryogenic potential of culture tissues (Kackar and Shekhawat 2007).

6.9.2.5 Polyamines as Breeders in Plant Symbiotic Associations on Economically Important Crops

Exogenous application of polyamines on microbial root associations leading to mycorrhiza or nodule formation is another phenomenon where this practice has been used. In fact, polyamines are considered a tool to improve cultivation (Wu and Zou 2009). The number of nodules in soybean (*Glycine max*) is affected negatively by foliar addition of spermidine and spermine (2 μM) (Terakado, Yoneyama, and Fujihara 2006). Nevertheless, mycorrhizal colonization presented a positive response to the addition of polyamines (100 mg/L) to the seedlings of citrus (*Citrus tangerine*) and orange (*Poncirus trifoliata*). Interestingly, different effects were observed depending on the type of polyamine; while orange responded to putrescine or spermine addition, spermidine had no effect over mycorrhizal development, and citrus responded to the addition of putrescine and spermine at level of the number of arbuscules, while spermidine and putrescine increased the number of vesicles (Wu and Zou 2009; Wu et al. 2010).

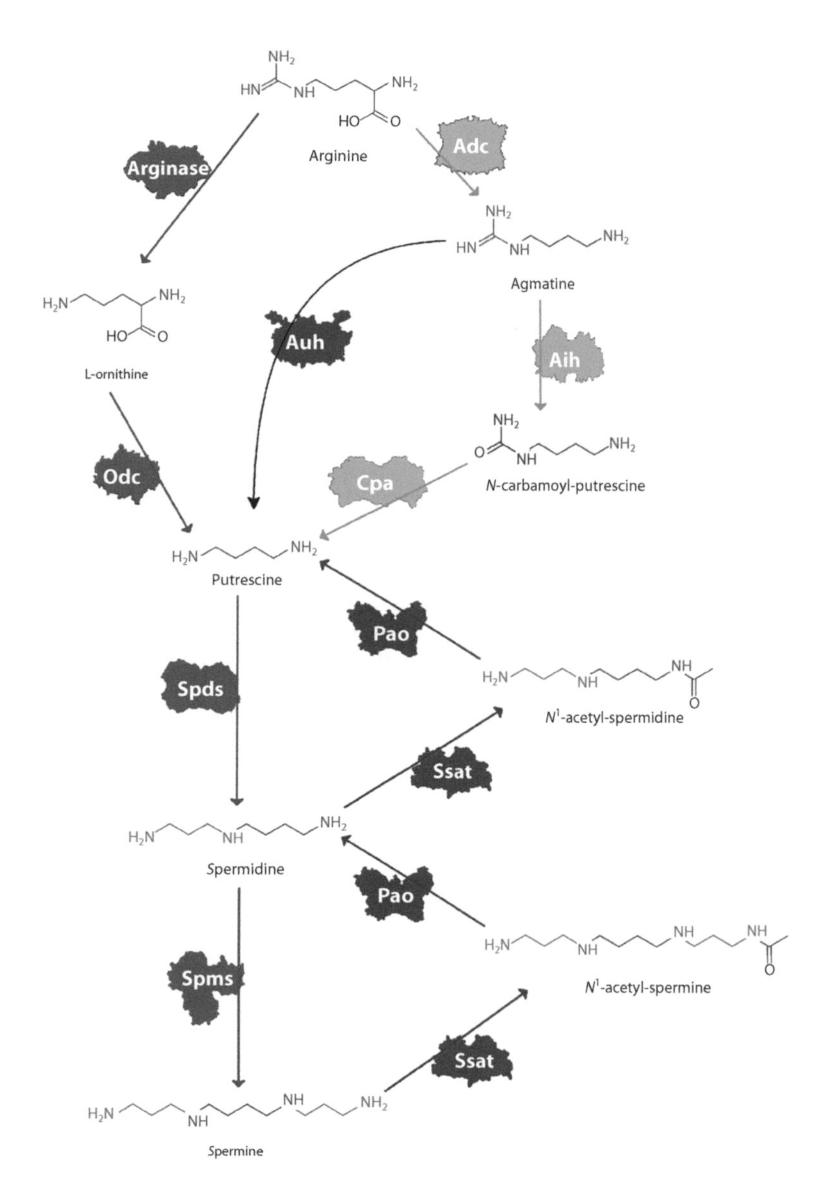

Figure 3.1 Polyamine biosynthetic and retroconversion pathways. Putrescine structure and/or 1,4-diamino-butanoyl groups are in red; amino-propyl groups are in green; the carbamoyl group is in magenta; imino groups are in blue; and carboxyl groups are in black. Enzymes are represented by cartoons, with white letters indicating their respective names. Enzymes involved in the biosynthetic polyamine pathway are in red; enzymes of the alternative putrescine biosynthesis starting with the Adc (arginine decarboxylase) are in green; those enzymes implicated on the back conversion pathway are in magenta; and the agmatine ureohydrolase (Auh) is in blue. Aih, agmatine iminohydrolase; Cpa, *N*-carbamoyl-putrescine amidohydrolase; Odc, ornithine decarboxylase; Pao, polyamine oxidase; Spds, spermidine synthase; Spms, spermine synthase; Ssat, spermine, spermidine N^1-acetyltransferase.

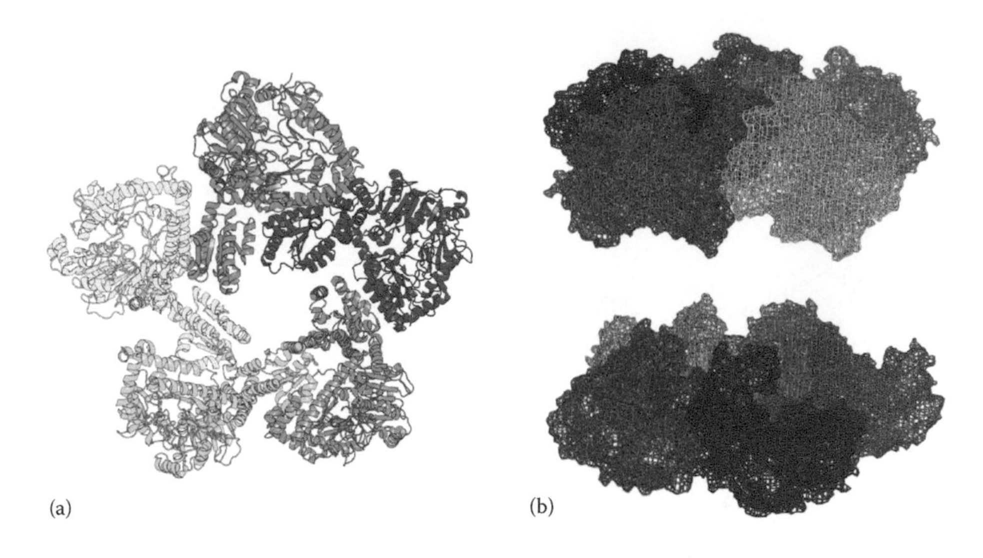

Figure 3.3 Structure of the acid-induced arginine decarboxylase (Adc) from *Escherichia coli*, 2VYC (Andréll, J., Hicks, M.G., Palmer, T., Carpenter, E.P., Iwata, S., Maher, M.J., *Biochem* 48, 3915–3927, 2009) created with Pymol (DeLano, W.L., *The PyMOL Molecular Graphics System*, DeLano Scientific LLC, San Carlos, CA, 2002; http://www.pymol.org). (a) Pentameric ring of Adc. (b) Full decameric structure of Adc represented by a mesh-surface visualization. Each monomer is highlighted with a different color.

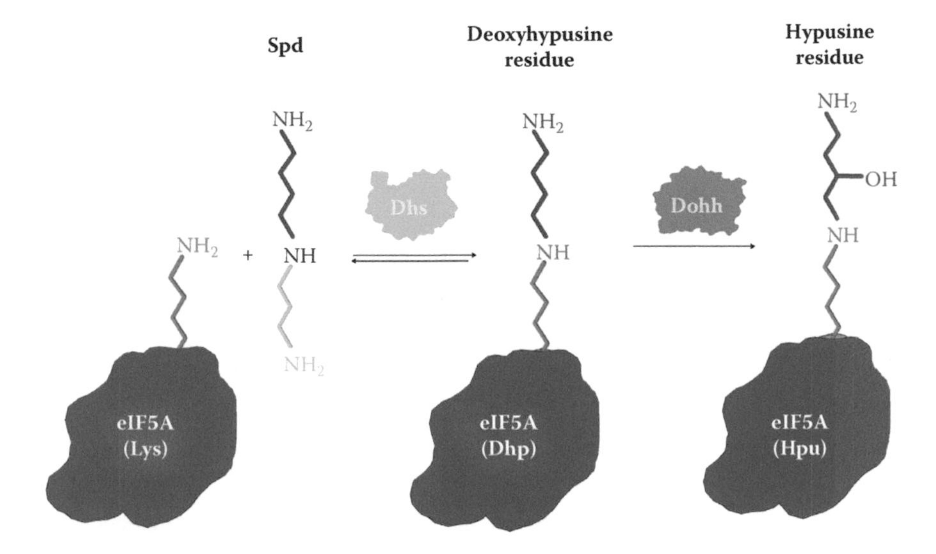

Figure 4.21 Schematic representation of the mechanism of activation of eIF5A. A modification in a specific lysine residue (blue) on eIF5A is performed by the transfer of an 4-aminobutyl moiety (red) from spermidine to lysine ε-amino group yielding Dhp, which is hydroxylated by the action of Dohh to form Hpu. Dhp, deoxyhypusine; Dhs, deoxyhypusine synthase; Dohh, deoxyhypusine hydrolase; Hpu, hypusine; Lys, lysine; Spd, spermidine.

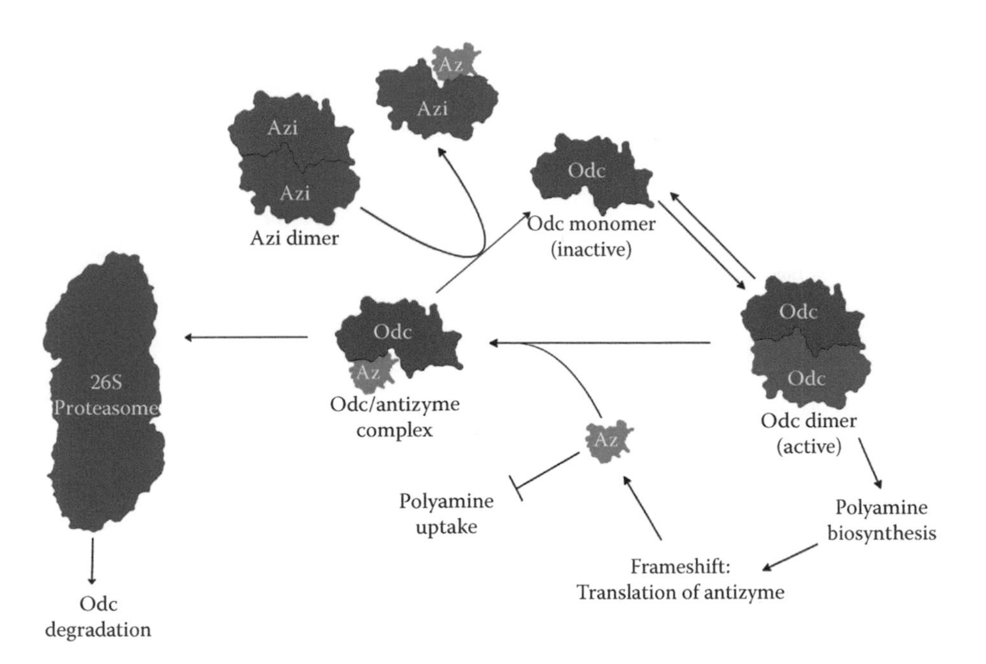

Figure 5.2 Schematic representation of the ornithine decarboxylase (Odc) regulation mediated by antizyme (Az). Odc homodimer is the active form. Az inhibits the formation of the Odc homodimer, leading Odc for degradation via the 26S proteasome without ubiquitination. At the same time, Az inhibits polyamine uptake. Polyamine biosynthesis is restored by displacement of Az by the Az inhibitor (AzI), allowing Odc homodimer formation.

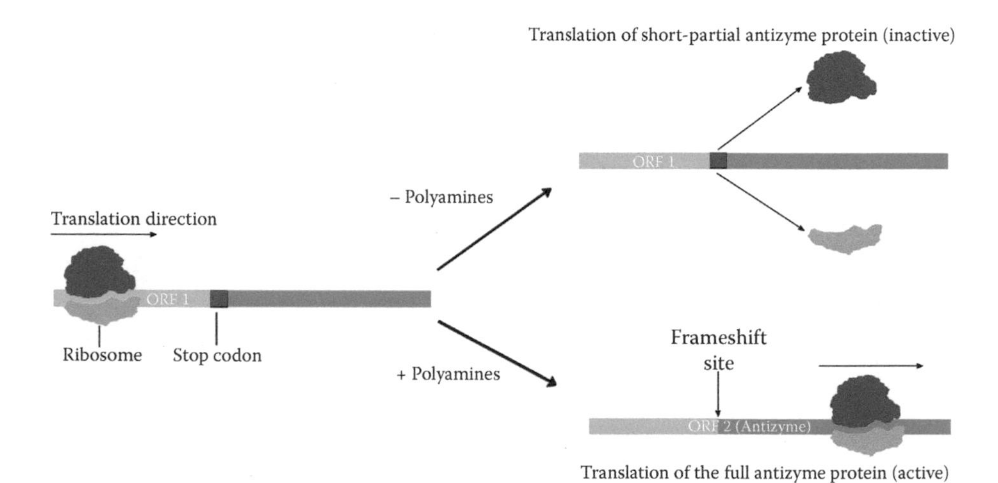

Figure 5.3 Schematic representation of the regulation of antizyme (Az) synthesis by polyamines. Az mRNA contains two overlapping open reading frames (ORFs). ORF1 contains a UGA stop codon that is read under conditions of low levels of polyamines, resulting in no functional Az. High levels of polyamines stimulate a +1 frame shift, allowing continuation of translation of ORF2, resulting in the synthesis of the complete and active Az form.

CHAPTER 7

The Role of Polyamines on Cell Differentiation and Morphogenesis in Fungi and Other Organisms

7.1 INTRODUCTION

Morphogenesis has received several definitions by different authors. Thus, Nickerson and Bartnicki-García (1964) define it as follows: "Morphogenesis encompasses those aspects of development related to morphological changes." "Morphogenesis refers to the development of the shape and form of the organism and its individual parts" was the definition of Garrod and Ashworth (1973), whereas Harold (1990) uses a slightly different definition: "morphogenesis will refer to the processes that generate the forms of cells in the course of their growth, division or development," and Moore (1998) indicates that morphogenesis "involves changes in the shape, form and/or structure of an organism as a result of a sequence of developmental arrangements." The same occurs with cell differentiation, which is defined by Nickerson and Bartnicki-García (1964) in these words: "differentiation connotes the acquisition of a determined function or structure during development"; Garrod and Ashworth (1973) use the definition "differentiation involves the structural and functional specialization of individual cells from one of a number of common basic stem cells which are usually competent to differentiate in several different ways"; Bennett (1983) indicates that differentiation is observed when an organism suffers a diversification of the structure and function of the cells or acquires differences during development; and Ruiz-Herrera (1984) uses a more mechanistic approach: "differentiation constitutes the series of events that carefully organized in time and space give rise to cell specialization, without alterations in their genetic characteristics." Independently of the definitions that better satisfies the readers, it is apparent that we have knowledge of these phenomena, although we cannot give a precise definition of them.

Polyamines have proved to be essential to carry out morphogenetic as well as differentiation processes. To understand the mechanisms by means of which polyamines regulate morphogenesis and differentiation, it is essential the use of simple models such as fungi. Furthermore, molecular genetic techniques such as microarrays are valuable tools in uncovering the mechanisms through which polyamines fulfill their many physiological actions (see Chapter 6). The present chapter will describe the existent knowledge regarding the role of polyamines as modulators of morphogenetic and differentiation processes, with emphasis on fungal systems.

As indicated in Chapters 2 through 4 of the volume, the polyamines most widely distributed are putrescine, spermidine, and spermine. These small ubiquitous polycationic molecules are synthesized in fungi starting from ornithine. The decarboxylation of ornithine by ornithine decarboxylase (Odc, EC 4.1.1.17) generates putrescine, the subsequent addition of an aminopropyl moiety to putrescine catalyzed by spermidine synthase (Spds, EC 2.5.1.16) produces spermidine, and a similar reaction but in this case using spermidine as substrate and catalyzed by spermine synthase (Spms, EC 2.5.1.22) produces spermine. In either case, the aminopropyl group transferred by Spds and Spms is donated by decarboxylated *S*-adenosylmethionine (dcSAM) generated by *S*-adenosylmethionine decarboxylase (Samdc, EC 4.1.1.50). The changes, both quantitative and qualitative in the cellular contents of polyamines, caused by alterations in growth conditions such as temperature, pH, and stressful environmental conditions or by internal physiological alterations, are regulated by complex mechanisms that have been described in previous chapters. Accordingly, we can come to the conclusion that the cellular polyamine contents, both quantitative and qualitative, depend on specific conditions at a determined stage of growth and development (Marshall et al 1979; Valdes-Santiago and Ruiz-Herrera 2014).

7.2 DIFFERENTIAL LOCATIONS OF POLYAMINES AND THEIR RELATION TO SPECIFIC PHENOMENA

In Chapter 2, we described the localization of polyamines in the cell. Here, we will describe the impact of the different locations of polyamines and their synthesizing enzymes in specific physiological processes and phenomena.

One important contribution of the *Mucor rouxii* model system in the study of polyamines was the demonstration of the existence of several Odc compartments, which may explain the different physiological processes regulated by the different natural polyamines (Martinez-Pacheco and Ruiz-Herrera 1993). It was demonstrated that diamino butanone (DAB), an inhibitor of Odc, had an inhibitory effect on spore germination, only at a specific phase of the process, just before germ-tube emergence (Ruiz-Herrera and Calvo-Mendez 1987). One possible explanation to this specificity was that DAB did not have access to the cells at another developmental stage because polyamines were located in another compartment not accessible to DAB. This hypothesis was supported when germlings were treated with diethylaminoethyl dextran (a nonpermeating high-molecular-mass compound that permeabilizes cells without injury to the integrity of cell organelles). By this treatment, Odc activity was reduced by DAB only to 50%, while control germlings treated with toluene-ethanol that destroy the selective permeability of the plasma and internal membranes suffered 100% inhibition by DAB (Martinez-Pacheco and Ruiz-Herrera 1993). This compartmentalization of Odc seems to be complex, as was shown in this system, since apparently, polyamines present in some compartments were involved in growth, and those present in other ones were related to cell differentiation (Martinez-Pacheco and Ruiz-Herrera 1993).

Not only Odc has been reported to be sequestered in specialized compartments, but also studies with Chinese hamster ovary (CHO) cells made with fluorescent polyamine probes suggested that polyamines enter into the cell via simple transport through the plasma membrane, are delivered in the cytoplasm, and later on enter into different compartments via other simple transporters (Soulet et al. 2002, 2004). Evidence showed that natural polyamines may behave in this same way; astrocytes from rat glial cells contain spermidine and spermine but they do not contain Odc, indicating that polyamines are produced in other cells such as neurons, and later on, they are released and sequestered and stored in vesicles which transport them to astrocytes. These results made the authors suggest that the intracellular distribution of polyamines is involved in different independent functions (Laube and Veh 1997; Laube et al. 2002). Antibodies against Spds did not recognize the same areas of its product in CHO cells, which supports the idea of internal delivery, uptake, and compartmentalization in structures present in different types of cells (Krauss et al. 2006). These data support the concept that the knowledge of the exact distribution of polyamines and the synthesizing enzymes is essential for the comprehension of their different modes of action.

7.3 THE USE OF FUNGI AS MODELS FOR THE STUDY OF THE MECHANISMS OF POLYAMINE ROLES ON CELL DIFFERENTIATION

Differentiation phenomena can be divided into several steps. First, the sensing of a specific external stimulus that will be processed through signal transduction is necessary. The response is observed as the expression of specific genes involved in the differentiation processes, and finally, the morphogenetic change takes place (Ruiz-Herrera and Hard 1997).

Fungi have evolved different and sophisticated strategies to complete their life cycles. Most of these strategies require morphogenetic events leading to the generation of various structures such as mycelium, spores, conidia, and appressoria (Cole 1986; Mendgen, Hahn, and Deising 1996). It has been demonstrated that polyamines are involved in all these processes and that variations in the contents of polyamines during the differentiation processes are related with alterations in morphology. Thus, it has been observed that certain morphological characteristics of fungi depend on their cellular polyamine composition and concentration; e.g., in *Paracoccidioides brasiliensis*, mycelial cells presented lower amounts of putrescine and spermidine than yeast cells did (San-Blas et al. 1997). Other dimorphic fungi such as *Ceratocystis ulmi*, *Ceratocystis fagacearum*, *Fusarium moniliforme*, and *Histoplasma capsulatum* have been reported to contain spermine in their yeast forms, whereas the mycelial forms do not contain this polyamine (Marshall et al. 1979). Another, example is *M. rouxii*, whose putrescine content was higher during yeast growth (Martinez-Pacheco et al. 1989). And we can mention that when exogenous putrescine was applied to the vesicular-arbuscular mycorrhiza fungus

Glomus caledonium, there occurred an increase in root colonization and biomass accumulation (Rutto et al. 1999).

It is known that *M. rouxii* spore germination is initiated when the environmental conditions are adequate. Thus, in the presence of oxygen, spores develop into mycelium, which eventually will produce sporangiophores, sporangia, and new spores. If oxygen is absent, and the atmosphere contains a high level of carbon dioxide, swollen spore growth occurs by budding in the form of yeasts (see Ruiz-Herrera 1994). However, as indicated in Section 7.2, if the Odc inhibitor DAB is added to spores, no germination takes place and the spores only round up and increase in size to reach large volumes (Ruiz-Herrera and Calvo-Mendez 1987). In addition, it is important to indicate that DAB treatment of germinating spores of *M. rouxii* not only inhibited their germination but also induced structural and chemical alterations in the cell wall. Thus, after 24 hours in the presence of the inhibitor, the cell wall increases about three times in thickness and shows alterations in its staining pattern. Also, it was observed that walls had lower levels of neutral sugars, proteins, and lipids compared with untreated spores, as well as a higher level of phosphate. At the morphological level, in DAB-treated cells, fragmentation of vacuoles, autophagocytic processes, and formation of abnormal nucleoli were observed (Obregon et al. 1990).

During phorogenesis (phorogenesis consists in the formation of specialized aerial hyphae that contain the spores) in different species of Mucorales, it was observed that, as occurred with spore germination of *M. rouxii* (see previous discussion), DAB inhibited phorogenesis only when it was added before the initiation of the process (Martinez-Pacheco et al. 1989). Similarly, it was observed that Samdc activity was not detected in ungerminated spores of *M. rouxii*; however, an exponential increase in Samdc activity was detected once germination started. 5 mM methylglyoxal(bis) guanyl-hydrazone (MGBG) inhibited Samdc activity together with fungal growth, and in the case of spores, MGBG affected spore germination only when the inhibitor was added during the first 2 hours of spore incubation; afterward, it was not able to stop germination (Calvo-Mendez and Ruiz-Herrera 1991). As was exposed previously, inhibitors of polyamine biosynthesis allowed the determination that polyamine effects are time dependent, suggesting that spermidine is required at the early stages of spore germination, while putrescine seems to be the polyamine important to initiate polarized growth during the later stages of spore germination (Calvo-Mendez and Ruiz-Herrera 1991). A report showing the correlation between polyamines, DNA methylation, and polarized growth in *M. rouxii* was discussed in Chapter 6 (Cano, Herrera-Estrella, and Ruiz-Herrera 1988).

All these data demonstrate that phenomena such as spore germination, phorogenesis, sporulation, mycelial growth, dimorphism, and also transition of mycelium to mature sclerotia and appressorium formation depend on polyamines, and it is well known that an increase in the concentration of polyamines precedes all morphogenetic events (Paulus, Kiyono, and Davis 1982; Rajam, Weinstein, and Galston 1986; Shapira et al. 1989; Balasundaram, Tabor, and Tabor 1993; Ruiz-Herrera 1994; Reitz, Walters, and Moerschbacher 1995; Khurana et al. 1996; Guevara-Olvera, Xoconostle-Cazares, and Ruiz-Herrera 1997; Herrero et al. 1999; Jimenez-Bremont, Ruiz-Herrera, and Dominguez 2001; Mueller et al. 2001; Valdes-Santiago,

Cervantes-Chavez, and Ruiz-Herrera 2009; Valdes-Santiago, Guzman-de-Pena, and Ruiz-Herrera 2010).

Also, as shown by the data described previously, the use of inhibitors of polyamine biosynthesis has been an important tool in the study of different aspects of the role of polyamines in cell physiology and in their possible use to control fungal infections. Difluoromethylarginine (DFMA) (Kallio, McCann, and Bey 1981; Rajam and Galston 1985), 1,4-diamino-2-butanone (DAB) (Stevens, McKinnon, and Winther 1977), difluoromethylornithine (DFMO) (Metcalf et al. 1978), and MGBG (Williams-Ashman and Scheone 1972; Pegg and Williams-Ashman 1987) are some of the most common polyamine biosynthesis inhibitors. DFMA is a specific inhibitor of arginine decarboxylase, DFMO and DAB are specific inhibitors of Odc, and MGBG inhibits Samdc. Besides the examples of the effects caused by the addition of some of the polyamine biosynthesis inhibitors cited previously, we cite a few more of them further in the chapter. Mycelial growth of *Ophiostoma ulmi* treated with DFMO, DFMA, or MGBG was reduced (Biondi, Polgrossi, and Bagni 1993). DFMO reduced spore germination and hyphal growth, while DFMA promoted hyphal growth in *Glomus mosseae* (El Ghachtouli et al. 1996). Something similar happened with *Laccaria proxima*, since DFMO slightly reduced growth but DFMA increased fungal growth (Zarb and Walters 1994). Exogenous application of MGBG reduced many differentiative processes in *Alternaria consortiale*, such as hyphal growth, as well as spore germination (Kepczynska 1995). Conidiation also was affected by the addition of DFMO, DFMA, and MGBG in *Aspergillum niger* (Khurana et al. 1996). Germ tube elongation and spore germination was affected in *Uromyces phaseoli* by DFMO, and additionally, a reduction in the percentage of the leaf area infected with *Vicia faba* was observed (Walters 1986; Rajam, Weinstein, and Galston 1989). MGBG inhibited conidial germination and appressorium formation in *Colletotrichum gloeosporioides* (Ahn et al. 2003). The phytopathogenic fungus *Pyrenophora avenae* growing with the addition of DFMO, MGBG, or ethylm ethylglyoxalbis(guanylhydrazone) was unable to form mycelium, and this effect was reverted by the addition of exogenous polyamines, only in the case when DFMO was added (Foster and Walters 1990).

Dimorphism in fungi has been defined as their ability to grow as mycelium or yeast-like, and this change occurs in either direction when induced by environmental conditions (Ruiz-Herrera and Hard 1997). It is important to stress that there are many examples of fungi that, no matter their taxonomic position, are dimorphic, and there are evidences showing that polyamines are essential for the dimorphic transition. Some of the first evidences of the role of polyamines in fungal dimorphism took place in Mucorales (Garcia et al. 1980; Inderlied, Cihlar, and Sypherd 1980; Cano-Canchola et al. 1992). In the case of *Mucor bacilliformis*, it was observed that some monomorphic mutants that grew constitutively in the yeast form contained very low levels of Odc activity (Ruiz-Herrera, Ruiz, and Lopez-Romero 1983). When the levels of polyamines and the activity of Odc were analyzed in two species of *Mucor*, *M. bacilliformis* and *M. rouxii*, during spore germination and dimorphism, it was observed that Odc activity was lower in the yeast form compared with the mycelium form. Odc activity was also affected by the age of cells, in such a way that young mycelium presented higher levels of Odc activity compared with old mycelium. Interestingly, dimorphism is reversible in *Mucor*, and it was observed that cells growing in the yeast form under anaerobic conditions with a high concentration of CO_2, when transferred to an aerobic or CO_2-free N_2 atmosphere, showed a rapid increase in Odc activity just before their change to the mycelium form. The opposite was observed with cells growing in the form of mycelium under aerobic conditions, which, when transferred to anaerobic conditions with a high concentration of CO_2, immediately displayed a reduction in Odc activity. This is an example that made establishing a correlation between Odc activity, morphogenesis, and polyamines possible (Calvo-Mendez, Martinez-Pacheco, and Ruiz-Herrera 1987). Interestingly, DAB, the competitive inhibitor of Odc, inhibits the transition from yeast to mycelium but not the opposite processes. The inhibition of dimorphism came along with a decline in the polyamine pools as well as in Odc activity. However, DAB did not affect the synthesis of polyamines to the point to inhibit cell growth. Both observations allowed the conclusion that certain levels of polyamines are required to sustain normal cell growth and higher levels of polyamines are obligatory during differentiation processes (Martinez-Pacheco et al. 1989).

The role of polyamine in dimorphism was further confirmed in many dimorphic fungi. Thus, it was reported that DAB reduced Odc activity and inhibited budding and the mycelium-to-yeast transition in *P. brasiliensis* and that dimorphism was accompanied by a high activity of Odc (San-Blas et al. 1996, 1997). Also, the yeast-to-mycelium transition was inhibited by DAB and DFMO in *Yarrowia lipolytica* (Guevara-Olvera, Calvo-Mendez, and Ruiz-Herrera 1993). The phytopathogenic fungus *Ustilago maydis* affected in the synthesis of Odc by a disruption in the gene encoding the enzyme, in contrast to the wild-type strain, was unable to form mycelium when dimorphism was induced by a pH change (Guevara-Olvera, Xoconostle-Cazares, and Ruiz-Herrera 1997). As could be expected, dimorphism ability of the fungus was recovered by the exogenous addition of high concentrations of putrescine (Guevara-Olvera, Xoconostle-Cazares, and Ruiz-Herrera 1997). Similar results were obtained when *odc* mutants of *Y. lipolytica* (Jimenez-Bremont, Ruiz-Herrera, and Dominguez 2001) and *Candida albicans* (Herrero et al. 1999) were similarly analyzed. It was therefore concluded that polyamines are required for the differentiation process of dimorphism in all fungal models (reviewed by Valdes-Santiago et al. 2012a).

7.4 ROLE OF POLYAMINES ON DIFFERENTIATION IN PLANTS

The essentiality of polyamines for morphogenesis and differentiation has been found also to be true for plants. Several studies have established an association between phenomena such as embryogenesis with elevated levels of polyamines (El Hadrami and D'Auzac 1992; Mala et al. 2009). Also, during peach floral development, the expression levels of *ADC* gene were increased (Liu and Moriguchi 2007).

As occurs in fungi, polyamine biosynthesis inhibitors have been useful in discovering polyamine physiological significance in plants. Floral defects were observed in tobacco cells culture, as well as in tomato (*Lycopersicon esculentum*) treated with the inhibitors MGBG or DFMO (Rastogi and Sawhney 1990; Trull, Holaway, and Malmberg 1992). Also, an incomplete inhibition of the growth of stems and leaves was reported in *Solanum tuberosum* when MGBG was added to the culture medium (Feray et al. 1994). Likewise, maturation of white spruce somatic embryos was inhibited by 54% and stimulated by 113% by the addition of MGBG or 5 mM spermidine, respectively (Meskaoui and Trembaly 2009). Other organisms such as the alga *Chattonella antiqua* have proved to be sensitive to MGBG treatments, since growth was suppressed by addition of the inhibitor, an effect that was partially reverted by addition of spermidine (Nishibori and Nishijima 2003).

Polyamines inhibit other physiological events such as flower senescence, probably by inhibition of ethylene synthesis, since polyamines and 1-aminocyclopropane, a precursor of ethylene, share *S*-adenosylmethionine as a common precursor (Roberts et al. 1984). In *Avena sativa* and *Vigna aconitifolia*, DFMA and DFMO reduced cell division and organogenesis, and this effect was reverted by polyamine addition (Sawhney, Shekhawat, and Galston 1985). Furthermore, it was reported that plastid development, as well as its metabolism, and the efficiency of photosynthesis were influenced by polyamines (Sobieszczuk-Nowicka and Legocka 2014).

7.5 USE OF FUNGAL AND PLANT MUTANTS IN THE ANALYSIS OF THE SPECIFIC ROLES OF THE DIFFERENT POLYAMINES IN CELL PHYSIOLOGY

A fruitful approach for studying polyamine functions has been the isolation of polyamine-deficient mutants. Mutants affected in the synthesis of polyamines have been isolated in several fungal systems, among them, *Spe* mutants of *Saccharomyces cerevisiae* unable to synthesize polyamines were originally isolated by Whitney and Morris (1978). Later on, Odc (*odc*) or Samdc (*samdc*) mutants were found to be unable to sporulate, and they showed increased sensitivity to oxygen (Balasundaram, Tabor, and Tabor 1993). *S. cerevisiae* mutants affected in the gene encoding Spds were short-lived, and this phenotype was reverted by spermidine as well as by putrescine addition. Likewise, the external administration of spermidine increased the chronological life span in wild-type cells, as well as the remaining replicative life span in old yeast cells (Minois, Carmona-Gutierrez, and Madeo 2011). Another example of the importance of mutants affected in polyamine biosynthesis to understand the role of these compounds in cell differentiation is *Spe-1* mutant of *Neurospora crassa*. This strain presented poor ascospore germination, as compared with the wild-type strain (Paulus, Kiyono, and Davis 1982).

Morphogenetic and development events in *U. maydis* mutants affected at different steps in polyamine metabolism, *odc*, *spds*, and *samdc*, were found to be, as expected, auxotrophic, requiring polyamine addition to grow. These mutants were also affected in multiple physiological characteristics such as response to ionic, osmotic, and oxidative stress, as well as dimorphic transition and virulence (see previous discussion, and Guevara-Olvera, Xoconostle-Cazares, and Ruiz-Herrera

1997; Valdes-Santiago, Cervantes-Chavez, and Ruiz-Herrera 2009; Valdes-Santiago, Guzman-de-Pena, and Ruiz-Herrera 2010; Valdes-Santiago et al. 2012a). As described previously, *Y. lipolytica odc* mutants behaved similarly to those treated with inhibitors of polyamine biosynthesis and were unable to carry out the dimorphic transition (Jimenez-Bremont, Ruiz-Herrera, and Dominguez 2001). Null *odc* mutant of *Tapesia yallundae* formed no infection plaque in forage grass, although their virulence was not affected (Mueller et al. 2001). On the other hand, an *odc* mutant from another plant pathogen, *Stagonospora nodorum*, showed attenuated virulence on wheat (Bailey, Mueller, and Bowyer 2000). *C. albicans* defective in the *ODC* gene was altered in hyphal development and showed increased sensitivity to osmotic stress and calcofluor white (Herrero et al. 1999).

Mutants in genes encoding enzymes involved in polyamine biosynthesis also have given information about the role of polyamines in plants. *Arabidopsis thaliana* affected in one of the two genes encoding Samdc presented a bushy and dwarf phenotype (Ge et al. 2006), and the *samdcs*, *adcs*, and *spds* double mutants were embryo lethal, indicating that polyamines are essential for embryogenesis (Urano et al. 2003; Imai et al. 2004a; Ge et al. 2006). Also, that polyamines are required for normal seed development in *Arabidopsis* was shown by the phenotype of *adcs* and *spds* double mutants (Imai et al. 2004a; Urano, Hobo, and Shinizaki 2005). In the same plant species, low levels of arginine decarboxylase activity induced abnormalities in root and in lateral growth (Watson et al. 1998).

As may be concluded by all the data described previously, the role of polyamine in morphogenesis, differentiation, and development is now well accepted, and no doubt that studies with fungi were crucial for arriving to this conclusion. Also, inhibitors of polyamine metabolism contributed enormously to this knowledge. However, side effects caused by inhibitors can cover up or even lead to confusing conclusions. In this regard, the use of mutants has allowed the confirmation of the roles of polyamine as it has been described in this chapter. Important for the comprehension of the mechanism of action of polyamines is the use of molecular genetic techniques such as microarrays as valuable tools for the identification of genes involved in the differentiation and morphogenesis processes regulated by polyamines, as indicated in Chapter 6.

CHAPTER 8

Polyamine Metabolism as a Target

8.1 INTRODUCTION

In previous chapters, we have described that, in general, all living beings require polyamines for their growth and development. It is no surprise, then, that synthesis of polyamines may be considered a target for the selective inhibition of growth or some of the adverse effects of different processes, such as virulence or cancer. For example, in plants, putrescine, the precursor of higher polyamines, is synthesized by direct decarboxylation of L-ornithine in a reaction catalyzed by ornithine decarboxylase (Odc) or by decarboxylation of L-arginine by arginine decarboxylase (Adc), leading to the formation of putrescine through agmatine and *N*-carbamoyl putrescine as intermediates. Accordingly, the alternative pathway (Adc) for the production of putrescine allows the selective inhibition of polyamine biosynthesis in plant pathogenic fungi by Odc inhibitors without affecting plant polyamine metabolism, taking into consideration that fungi utilize only the Odc pathway for putrescine biosynthesis (Valdes-Santiago et al. 2012a). Suicide Odc inhibitors such as difluoromethylornithine (DFMO) and a potent inhibitor of *S*-adenosylmethionine decarboxylase (Samdc), methylglyoxalbis(guanylhydrazone) (MGBG), would be the best inhibitors available to selectively block polyamine biosynthesis.

Polyamine catabolism is achieved by the spermidine/spermine-N^1-acetyltransferase (Ssat), which transforms spermine and spermidine, respectively, into N^1-acetylspermine and N^1-acetylspermidine, which are substrates for the polyamine oxidase (Pao) enzyme to generate spermidine and putrescine, respectively (Seiler 2004). Although few specific inhibitors of these reactions have been described, the enzymes involved are also possible targets as described further in Section 8.2.

In this chapter, how the inhibitors of polyamine biosynthesis shown in Table 8.1, and whose structure is shown in Figure 8.1, have proved to be very effective in retarding the growth and, in some cases, differentiation processes of several phytopathogenic fungi is described. Furthermore, it is important to stress that polyamine metabolism as target for infections is not confined only to phytophatogenic fungi, rather it can also be useful to the control of tumor growth, human pathogenic fungi, and parasitic infections in humans, as it is also discussed in the chapter.

Table 8.1 Fungal Species That Have Been Treated with Different Inhibitors of Polyamine Metabolism

Fungal Species	Inhibitor(s) Used	Reference
Colletotricum gloesporioides	MGBG	Ahn et al. 2003
Septoria tritici	DFMO	Smith, Barker, and Jung 1990
Ustilago maydis	DFMO	Smith, Barker, and Jung 1990
Septoria nodorum	DFMO	Bailey, Mueller, and Bowyer 2000
Pyrenophora avenae	DFMO	Bailey, Mueller, and Bowyer 2000
Ophiostoma ulmi	DFMO	Bailey, Mueller, and Bowyer 2000
Sclerotinia sclerotiorum	DFMO, MGBG	Gamarnik, Frydman, and Barreto 1994; Chibucos and Morris 2006
Phytophthora sojae	DFMO	Chibucos and Morris 2006
Colletotrichum truncatum	DFMO	Gamarnik, Frydman, and Barreto 1994
Rhizoctonia solani	DFMO	Walters 1995
Botrytis cinerea	DFMO	Rajam and Galston 1985; West and Walters 1989; Mackintosh and Walters 1997
Fusarium oxysporum	DFMO	Rajam and Galston 1985; West and Walters 1989; Mackintosh and Walters 1997
Cochliobolus carbonum	DFMO	Rajam and Galston 1985; West and Walters 1989; Mackintosh and Walters 1997
Phytophthora infestans	DFMO, MGBG	Walters 1995
Botrytis cinerea	DFMO	Walters 1995
Helminthosporium maydis	DFMO	Birecka et al. 1986
Uromyces viciae-fabae	DFMO, MGBG	Reitz, Walters, and Moerschbacher 1995
Helminthosporium oryzae	DFMO, DFMA, MGBG	Rajam and Rajam 1996
Curvularia lunata	DFMO, DFMA, MGBG	Rajam and Rajam 1996
Pythium aphanidermatum	DFMO, DFMA, MGBG	Rajam and Rajam 1996
Colletotrichum capsici	DFMO, DFMA, MGBG	Rajam and Rajam 1996
Gigaspora rosea	DFMO	Sannazzaro et al. 2004
Mucor rouxxi	DAB, DFMO	Calvo-Mendez, Martinez-Pacheco, and Ruiz-Herrera 1987; Martinez-Pacheco and Ruiz-Herrera 1993
Yarrowia lipolytica	DAB, DFMO	Guevara-Olvera, Calvo-Mendez, and Ruiz-Herrera 1993
Aspergillus nidulans	DAB	Stevens, McKinnon, and Winther 1977
Criptococcus neoformans	DFMO, Cyclohexylamine	Pfaller, Riley, and Gerarden 1990

Source: Modified from Valdes-Santiago, L., Cervantes-Chavez, J.A., Leon-Ramirez, C.G., Ruiz-Herrera, J., *J Amino Acids*, 2012, 1–13, 2012.

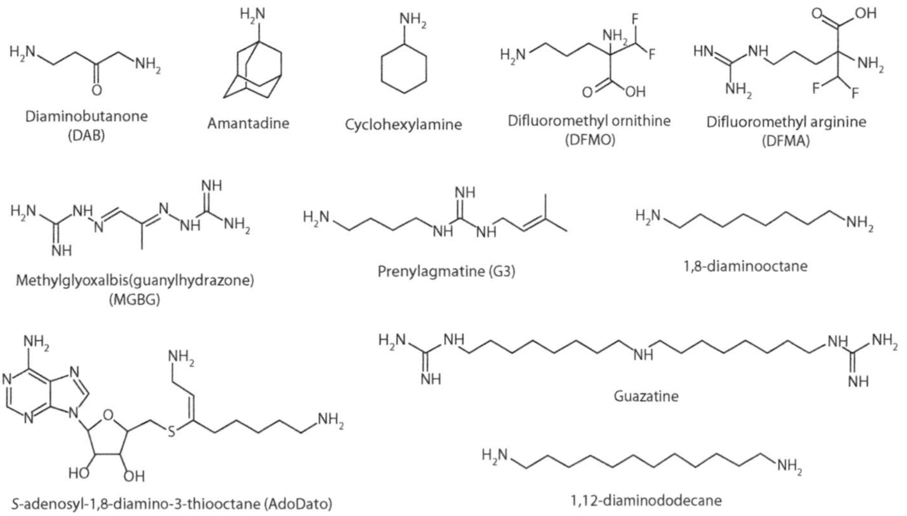

Figure 8.1 Structures of some inhibitors of polyamine metabolism.

8.2 EFFECT OF INHIBITORS ON THE DIFFERENT ENZYMES INVOLVED IN POLYAMINE METABOLISM

Inhibitors of the enzymes of the polyamine biosynthetic route have been considered extremely useful, not only for the study of polyamine physiological roles, as was described in Chapter 7, but also for the control of important human diseases such as cancer and to treat plant diseases by phytopathogenice fungi (Bartholeyns 1983; Oredsson, Friend, and Marton 1983; Demey 1987; Rosi et al. 2012). Regarding the mode of action of the inhibitors, it may be relevant to indicate that some of them are extremely specific, whereas others may act at different levels and when used in therapeutic applications may have side effects that make their use somewhat erratic. In Table 8.1, we show the data on the mode of action of the best known inhibitors, and in the following text, we extend this aspect with specific examples.

As mentioned in Section 8.1, probably the best studied inhibitors correspond to the first enzyme of the pathway of polyamine biosynthesis, Odc. α-DFMO has been widely used because of its specificity and its action as an irreversible inhibitor of Odc (Metcalf et al. 1978). DFMO specificity has led since the beginning in its use as a curative agent, e.g., in cancer (reviewed by Meyskens and Gerner 1999), in the treatment of sleeping disease produced by *Trypanosoma brucei* (Van Nieuwenhove et al. 1985), and in the treatment of the human cutaneous and mucocutaneous leishmaniases caused by *Leishmania mexicana* (Gonzalez et al. 1991). Structural analysis in the Odc enzyme of *Entamoeba histolytica* (EhOdc) showed that some substitutions in the active site provided EhOdc resistance to DFMO; this resistance has been explained as a presumable adaptive mechanism of the pathogen (Jhingran et al. 2008; Preeti et al. 2013). Another inhibitor specific for Odc is diaminobutanone (DAB), although this substance has been used mostly in the laboratory to determine the effects of Odc inhibition, and its use in therapeutics has been centered only in the treatment of some parasitosis (Reis et al. 1999; Vannier-Santos et al. 2008).

At the structural level, the similarity of MGBG to spermidine has made it a good candidate to act as a competitive inhibitor of *S*-adenosylmethionine (SAM) decarboxylase (Holtta et al. 1973). However, it has been reported that MGBG might affect mitochondrial integrity, maybe by destabilization of its membrane structure, a problem that has to take into consideration when used in therapeutics. Likewise, MGBG prevents spermine binding to the mitochondrial membrane and spermine entry into the cell (Oredsson, Friend, and Marton 1983; Toninello et al. 1999). MGBG has been used to control diseases such as psoriasis, since topical application inhibited epidermal DNA synthesis associated to modifications in polyamine metabolism (McCullough et al. 1983). In contrast, MGBG was not efficient in the treatment of myeloma (Winter et al. 1990). Patients with leukemia have been treated with MGBG, despite that severe side effects were presented (Rosenblum et al. 1981; Gastaut et al. 1987) and positive response has been observed in patients with AIDS-associated non-Hodgkin lymphoma (Von Hoff 1994).

In the case of inhibitors of spermidine synthase (Spds), it was found that most of them also inhibited spermine synthase (Spms) because of the similarity of Spds with Spms. Probably the most specific inhibitor of the enzyme is *S*-adenosyl-1,8-di-amino-3-thiooctane, which inhibited Spds from different organisms. Another inhibitor of Spds is Cyclohexylamine, which has been described as a potent inhibitor of Spds from several sources, although the enzymes from some bacteria were resistant to inhibition. However, despite their activity *in vitro*, no reports exist of the use of these inhibitors as therapeutic agents (Pegg 1987).

The inhibitors of Spms also have not been used in clinical trials, and their effect on the enzyme has been tested also in *in vitro* experiments (Shirahata et al. 1993).

Some inhibitors of polyamine catabolism have been also described as active against some diseases (Fogel-Petrovic et al. 1993a), among them amantadine. This compound contains an achiral amino group that is acetylated by spermine/spermidine-N^1-acetyltransferase (Ssat), acting as a competitive inhibitor of the enzyme. Studies in transgenic mice overexpressing Ssat suggested that amantadine is acetylated exclusively by Ssat, explaining its specificity (Aoki and Sitar 1988; Bras et al. 1998, 2001). Amantadine also has antiviral properties and has been used against influenza A and, interestingly, in the treatment of Parkinson's disease. Nevertheless, the mechanism of action of amantadine as antiviral and its use for the treatment of Parkinson's disease have not been attributed to its activity as an Ssat inhibitor. It has been demonstrated that amantadine binds to the phospholipid bilayer of M2, a membrane proton channel protein of influenza A virus (Cady et al. 2009), whereas in the treatment of Parkinson's disease, amantadine acts as an antagonist of *N*-methyl-D-aspartate (Paquette et al. 2012). It is noteworthy that the addition of 5 mM putrescine did not revert the phenotype, which can be explained by a possible blocking of polyamine uptake by amantadine owing to its ability to bind to phospholipid bilayers.

Inhibitors of Pao have been also developed with the objectives to study them for use both in polyamine physiology and as antiproliferative drugs. Among them, we can cite specific inhibitors such as 1, 12-diaminododecane and *N*-prenylagmatine (G3) (Bianchi et al. 2006). Maize Pao protein has been used as a model for the analysis of the mechanism of action of Pao inhibitors. It was determined that the orientation of inhibitors in their interaction with the catalytic tunnel of the enzyme has an essential role in defining the inhibitory activity (Cona et al. 2004). Most organisms express Pao in a constitutive way; however, the specific inhibitor N^1, N^{11}-bis(ethyl) norspermine increased Pao expression in a cell lung carcinoma line (Wang et al. 2001b).

In our laboratory, we have studied the effect of two inhibitors of polyamine catabolism, amantadine, a competitive inhibitor of Ssat, and 1, 12-diaminododecane, a specific inhibitor of Pao, on the growth of the phytopathogenic fungus *Ustilago maydis* single and double polyamine mutants (81 [*odc/spds*], LG4 [*odc*], LV54 [*spds*]) compared with wild type (FB2). All of the mutants presented auxotrophy to polyamines, either putrescine or spermidine (Guevara-Olvera, Xoconostle-Cazares, and Ruiz-Herrera 1997; Valdes-Santiago, Cervantes-Chavez, and Ruiz-Herrera 2009). When 5 mM amantadine was added to the media containing 0.5 mM spermidine to sustain growth, the *odc/spds* double mutant was the most affected in growth, whereas the *odc* mutant was less affected. In contrast, growth of the wild-type strain and the *spds* mutant did not present significant phenotypic changes (Figure 8.2). The

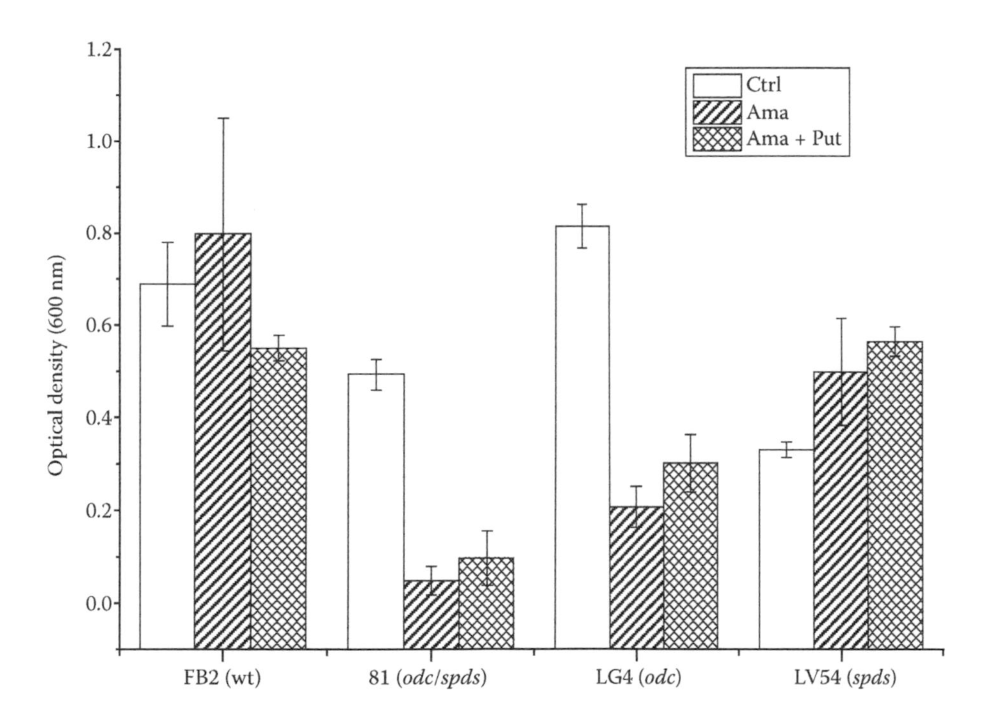

Figure 8.2 Effect of 5 mM amantadine, an inhibitor of Ssat, on the growth of *U. maydis* wild type strain and several mutants affected in genes involved in polyamine biosynthesis. Where indicated 5 mM putrescine was added. Ama, amantadine; Ctrl, control without additions.

possible explanation for these results is that inhibition of Ssat in *odc/spds* and *odc* mutants would not allow the synthesis of putrescine, in opposition with *spds* mutants that are able to produce putrescine directly by the decarboxylation of ornithine. These results were interpreted considering that putrescine has a role in growth, although a null mutant unable to generate putrescine because of an interruption in both *Odc* and *Pao* genes was able to grow only with the addition of spermidine (Valdes-Santiago, Guzman-de-Pena, and Ruiz-Herrera 2010). Accordingly, an alternative explanation would be that the location of polyamines in different cellular pools is an important factor to permit or not the access of the inhibitor to the enzyme. On the other hand, the addition of 0.5 μM 1,12-diaminododecane, a Pao inhibitor, caused a severe effect on the growth of *odc/spds*, *odc*, and *spds U. maydis* mutants, as well as the wild type. This effect was reverted by the addition of 5 mM putrescine (Figure 8.3). The structural hydrophobic features of 1,12-diaminododecane and its flexibility allowed it to enter into the cells (Weiger, Langer, and Herman 1998). As shown in Figure 8.3, the major effect was observed in the double mutant *odc/spds*. Two possible explanations for these results can be mentioned. It has been observed that polyamine analogs provoke multiple effects, not only down-regulation of the activity of the target enzyme but also a decline in both polyamine uptake and excretion (Wallace and Fraser 2004).

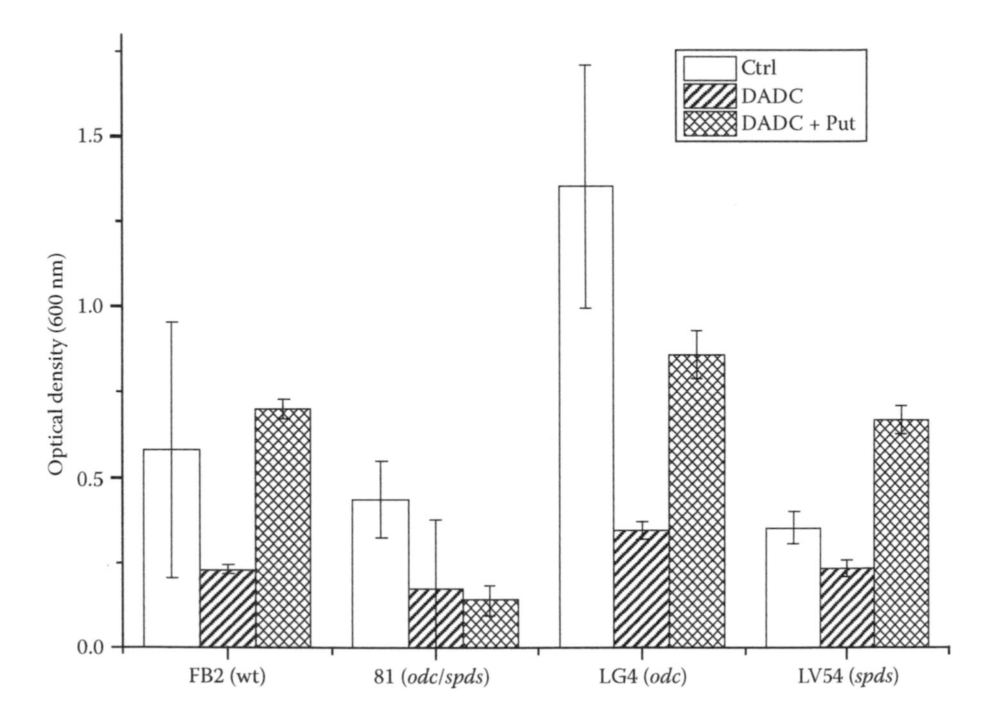

Figure 8.3 Effect of 0.5 μM 1, 12-diaminododecane (DADC), an inhibitor of Pao, on the growth of *U. maydis* wild-type strain and several mutants affected in genes involved in polyamine biosynthesis. Where indicated 5 mM putrescine was added. Ctrl, control without additions; Put, putrescine.

The absence of both genes encoding Odc and Spds may be responsible for the major metabolic changes compared with the changes induced in the single mutant *spds*, or the severity of the double mutation could produce an additional requirement of both putrescine and spermidine, not covered by the supplemented extracellular spermidine. This study is an example of the use of inhibitors and mutants for the comprehension of the roles of polyamines in cell physiology.

8.3 POLYAMINE METABOLISM AS A POSSIBLE TARGET FOR THE CONTROL OF SOME HUMAN DISEASES

It is well documented, and it has been discussed in previous chapters, that polyamines are essential for growth and that high concentrations of polyamines are necessary to carry out differentiation processes in all living organism. *De novo* synthesis, degradation, and transport pathways control intracellular concentration of polyamines; accordingly, alterations in either of these processes will bring about problems for any organism, and therefore, as pointed out, the substances that inhibit these processes have been used in the control of different diseases. Regarding human pathogenic fungi, data obtained *in vitro* have given credence to the possible use of the inhibition of polyamine metabolism for the control of the diseases that they inflict in humans. Accordingly, it was observed that an *spds* minus mutant of the thermally dimorphic fungus *Penicillium marneffei*, a pathogen of immunocompromised patients, showed defects in conidiogenesis, germination, and growth. These results led to the suggestion that the spermidine biosynthetic pathway might serve as a potential target for combating its infections (Kummasook et al. 2013). *Cryptococcus neoformans* is a facultative intracellular pathogen that causes severe meningoencephalitis or disseminated mycosis in immunocompromised patients (Perfect et al. 1998; Feldmesser et al. 2000); accordingly, cryptococcal meningitis is a major cause of death among HIV-infected individuals worldwide (Park et al. 2009). It is also a potentially deadly concern for cytotoxic chemotherapy for malignances and organ transplant patients (Mitchell and Perfect 1995; Mitchell et al. 1995; Kozubowski and Heitman 2012). It has been reported that DFMO and cyclohexylamine (an inhibitor of Spds) are able to inhibit the *in vitro* growth of *C. neoformans* (Pfaller, Riley, and Gerarden 1990). More recently, it was found that the *SPDS* gene was part of a chimera with the *LYS9* gene encoding saccharopine dehydrogenase, an enzyme involved in lysine biosynthesis. *In vitro* phenotypes of *spe-LYS9* mutants included reduced capsule and melanin production and growth rate, besides being avirulent (Kingsbury et al. 2004). Another opportunistic pathogenic fungus, *Aspergillus nidulans*, which causes fatal invasive aspergillosis in immunocompromised patients, was tested by the RNA interference (RNAi) strategy to cause a specific a silencing effect of the *Odc* gene *in vitro*, germinating spores, leading to a significant reduction in mycelial growth (Khatri and Rajam 2007). A further observation was that *odc* mutants of *Candida albicans*, considered the most important fungal human pathogen, grew exclusively in the yeast form at low polyamine levels (Herrero et al. 1999). However, *C. albicans odc/odc* (the fungus is diploid) mutants were virulent in a mouse model (Herrero et al. 1999).

The demonstration that the use of DFMO controlled *T. brucei* in humans without apparent harm to the host and that it is considered the safest and most effective drug in the treatment of human African trypanosomiasis (Bacchi et al. 1983; Heby, Roberts, and Ullman 2003) shows that inhibition of polyamine metabolism can be a control method of protozoan (and fungal) diseases, but as illustrated by the few examples cited in this chapter, sensitivity to polyamine inhibitor such as DFMO varies widely. Hence, the use of alternative inhibitors of polyamine metabolism is imperative. Accordingly, several polyamine analogs have been proposed to control fungi (Merali et al. 2000; Wallace and Fraser 2003; Wallace and Niiranen 2007), although the effect of these polyamine analogs on pathogenic fungi needs to be analyzed as an alternative to control fungal diseases.

In humans, several diseases have been related with alterations of polyamine metabolism. Patients with different types of cancer have presented high levels of urinary polyamines, and after chemotherapy treatment, a decrease in the levels of these substances was observed (Durie, Salmon, and Russell 1977; Wallace and Fraser 2004; Minois, Carmona-Gutierrez, and Madeo 2011). Also, alterations in polyamine metabolism have been related to the physiology of cardiac cells, a reason polyamines have been considered a target in the treatment of some cardiac diseases (Cetrullo et al. 2010, 2011; Giordano et al. 2010). A decrease in the expression of Ssat in the regions of the brain related with Parkinson's disease, together with others findings, such as a relationship between polyamines and the increased aggregation of α-synuclein, a protein connected in a genetic way with the disease, suggested a link between polyamines and the development of Parkinson's disease (Antony et al. 2003; Lewandowski et al. 2010; Stefanis 2012).

8.4 USE OF THE DIFFERENCES EXISTING IN THE MECHANISM OF POLYAMINE BIOSYNTHESIS BETWEEN PLANTS AND FUNGI AS A MECHANISM FOR THE CONTROL OF PHYTOPATHOGENIC FUNGI

The metabolism of polyamines in infected hosts suffers profound alterations in response to the pathogen attack. Interestingly, the pattern of changes in fungal infections depends upon whether the fungus is a biotroph or a necrotroph parasite (Walters 2000). Even during symbiotic relationships, such as in arbuscular mycorrhiza, a change in polyamine metabolism is a frequent response (Smith and Read 1997).

The idea that polyamines can be a target for the control of phytopathogenic fungi was born from the fact that, as indicated previously, plants possess two pathways for putrescine generation, whereas in opposition, most fungi contain only one route leading to putrescine biosynthesis. This feature makes the polyamine biosynthetic pathway suitable to be a target for the control of fungal diseases in plants (reviewed by Valdes-Santiago et al. 2012a).

As was indicated in previous chapters, plants generate putrescine either from arginine by the action of Adc plus two additional steps (Watson et al. 1998) or from ornithine by Odc, with the latter being the only biosynthetic route used by fungi.

The rest of the steps required for the generation of spermidine and spermine (in the organisms that posses this latter polyamine) are common to plants and fungi. Accordingly, these biosynthetic steps involve the addition of an aminopropyl group donated by decarboxylated SAM (dcSAM). The decarboxylation of SAM is carried out by the action of SAM decarboxylase (Samdc). Once the aminopropyl group is transferred to putrescine by the enzyme Spds, spermidine is formed. Using the same mechanism, Spms transfers another aminopropyl group to spermidine to form spermine (Pegg and McCann 1982).

The control of fungal pathogens requires a deep comprehension of the intricate metabolic interactions that take place in plants, where, as indicated previously, polyamine metabolism represents a good option to be used as a target. Taking into consideration the only difference existing in the biosynthetic route of polyamines indicated previously, most of the work done to control plant diseases caused by fungi using this route as target has been focused on the inhibition of Odc (Table 8.1). This is because plants are un affected by ODC inhibitors (reviewed by Valdes-Santiago et al. 2012a). Inhibitors such as DFMO, which binds irreversibly to the rate-limiting enzyme Odc, and DAB, a competitive inhibitor of ODC that acts as an analog of the product spermidine, have been widely used (Rajam and Galston 1985; Slocum and Galston 1985; Birecka et al. 1986; West and Walters 1989; Smith, Barker, and Jung 1990; Walters 1995; Kumar et al. 1997; Mackintosh and Walters 1997; Pieckenstain et al. 2001). An example of these studies is the observation that levels of 5 mM DFMO seem to be inoffensive to *Triticuma estivum*, supporting the use of such inhibitor to protect the plant against phytopathogenic fungi (Bharti and Rajam 1995). DFMO and DFMA (DL-α-difluoromethyl arginine; an inhibitor of Adc) have proved to be useful to inhibit the growth of *Botrytis cinerea* and *Penicillium expansum*, causal agents of soft rot in apple (Saftner and Conway 1997).

Nevertheless, some of the problems presented in the use of Odc inhibitors are that some stages in the fungal life cycle are unaffected by them; thus, while DFMO inhibited mycelial growth of *Sclerotinia sclerotiorum*, it was unable to inhibit ascospore germination (Pieckenstain et al. 2001). This difference could be explained by the low degree of entry or permeability of DFMO to the ascospores, compared with the mycelium (Garriz et al. 2003). Other consideration regarding the inhibition of putrescine of the fungal pathogens is the ability of fungi to obtain the polyamines they require from their hosts (Garriz et al. 2003). In this regard, combination of DFMO with polyamine transport inhibitors can be an option for the control of pathogens (Samal et al. 2013; Muth et al. 2014).

Probably because of the problems mentioned previously, DFMO has not proved to be useful in the case of some plant pathogens such as *Septoria tritici*, *U. maydis*, *Stagonospora nodorum*, *Pyrenophora avenae*, and *Ophiostoma ulmi* (Smith, Barker, and Owen 1992; Bailey, Mueller, and Bowyer 2000). Samdc also has been used as a target to control fungal plant diseases, as is the case of infections by *Colletotrichum gloeosporioides*, a pathogen of red pepper and avocado; MGBG is an Samdc inhibitor that presented effects over the germination of conidia and appressorium formation (Ahn et al. 2003); nevertheless, in the rice pathogen *Magnaporthe grisea*, neither DFMO nor MGBG affected appressorium function and/or development

(Choi et al. 1998). Some fungi such as *Gigaspora rosea*, *Colletotrichum truncatum*, *O. ulmi*, and *Laccaria proxima* are some examples of fungi containing Adc, and therefore, they are not suitable to be tested with DFMO (Khan and Minocha 1989; Biondi, Polgrossi, and Bagni 1993; Gamarnik, Frydman, and Barreto 1994; Zarb and Walters 1994; Sannazzaro et al. 2004). A list of fungi in which the effects of polyamine synthesis inhibitors have been analyzed is presented in Table 8.1 (reviewed by Valdes-Santiago et al. 2012a).

8.5 RNA INTERFERENCE AS A TOOL TO ANALYZE POLYAMINE METABOLISM AND TO CONTROL PATHOGENS BY SELECTIVE INHIBITION OF POLYAMINE BIOSYNTHESIS

RNAi is an efficient posttranscriptional mechanism for gene silencing, consisting of the activation of a sequence-specific RNA degradation process (Fere et al. 1998; Agrawal et al. 2003). RNAi is triggered by the presence of an exogenous double-stranded RNA (dsRNA) in the cell. dsRNAs are processed to generate small non-coding RNAs (sRNAs) of about 22–30 nucleotides by Dicer-2, an RNase III enzyme (Bernstein et al. 2001; Ketting et al. 2001). The RNA-induced silencing complex (RISC), together with argonaute-2 protein, processes sRNA and is used to guide RISC to the target RNAs, cleaving the mRNA by its endonuclease activity (Hammond et al. 2000, 2001; Meister 2013).

Beyond of the biological relevance of RNAi's role as a defensive pathway against viruses, repetitive genes, and transposons (Cullen 2002; Chang, Zhang, and Liu 2012), RNAi has emerged as a novel tool to explore the genes function, to control the biosynthesis of unwanted substances such as toxins from plants, identification of genes of interest, as well as the generation of crop plants with nematode resistance (Clemens et al. 2000; Fraser et al. 2000; Ogita et al. 2003; Kim et al. 2005; Guggenberger, Ilgen, and Adamskim 2007; Tamilarasan and Rajam 2013). In addition, it must be indicated that RNAi is relatively simple to implement (Ramadan et al. 2007; Bose and Doering 2011; Fellmann and Lowe 2014).

Examination of polyamine metabolism has been approached by use of RNAi techniques. Synthetic posttranscriptional control of Ssat by RNAi was used to demonstrate that induction of apoptosis by the analog N^1, N^{11}-diethylnorspermine (DENSPM) was directly related with the induction of both *Ssat* mRNA by DENSPM and Ssat activity. Spermidine and spermine pools induced by DENSPM were found to be inhibited by small interfering Ssat (Pegg, Pakala, and Bergeron 1990; Fogel-Petrovic et al. 1993b; Chen et al. 2003). Regulation of polyamine biosynthesis also has been investigated with the help of RNAi in trypanosomes, since regulatory mechanisms are not the same in mammalian cells. In this system, no antizyme homolog has been identified, and Odc and *S*-adenosylmethionine decarboxylase seem not to be subjected to a high turnover (Phillips, Coffino, and Wang 1987; Ghoda et al. 1990; Willert, Fitzpatrick, and Phillips 2007). However, gene silencing of *Odc*, *Spd*, and *Samdc* genes by RNAi induced an increase in an Samdc inactive homologue that forms a heterodimer with Samdc, activating the enzyme. Translational regulation was suggested when it was observed that Samdc RNAi increased the levels of a Samdc inactive homolog, as well as Odc (Willert and Phillips 2008; Xiao, McCloskey, and Phillips 2009).

The feasibility of this technique in the case of fungal diseases is illustrated by the first report of RNAi effectiveness in a fungal system using *C. neoformans*. One of the constructions to induce silencing consisted of a hairpin formation of 500 bp of the *CAP59* gene in opposite orientation, containing between them a set of 250 bp of the green fluorescent protein sequence. This construction was used to transform protoplast of *C. neoformans*. The *CAP59* gene is involved in the formation of the capsule in *Cryptococcus*, which is essential for virulence. It was observed that the transformed strain, CAP57i, presented the same phenotype compared with mutants deficient in the same gene obtained by conventional techniques (Liu et al. 2002). An analysis of the potential of this technique for the control of pathogenic fungi has been reviewed by Moazeni et al. (2014).

Now, regarding the use of genes encoding enzymes involved in polyamine metabolism as a target for fungal control, the first report was a study showing that *A. nidulans* germinating spores were able to uptake siRNA and incorporate it into the cells causing specific silencing effect of the gene encoding Odc (Khatri and Rajam 2007). A demonstration that the technique is effective in the control of a plant disease caused by a fungus, by silencing a gene involved in polyamine metabolism, was reported by Rajam (2012). This author transformed tobacco plants with an antisense construction of the mouse *ODC* gene and demonstrated that they were partially resistant to *Verticillium* wilt.

References

Abdulhussein, A. A. and H. M. Wallace. 2014. Polyamines and membrane transporters. *Amino Acids* 46: 655–660.

Adachi, M. S., J. M. Torres, and P. F. Fitzpatrick. 2010. Mechanistic studies of the yeast polyamine oxidase Fms1: Kinetic mechanism, substrate specificity, and pH dependence. *Biochemistry* 49: 10440–10448.

Agostinelli, E. 2012. Spermine oxidation products induce mitochondrial alteration on tumor. *Cell Scient J Fac Med Nis* 29: 111–116.

Agostinelli, E., G. Arancia, L. D. Vedova, F. Belli, M. Marra, M. Salvi, and A. Toninello. 2004. The biological functions of polyamine oxidation products by amine oxidases: Perspectives of clinical applications. *Amino Acids* 27: 347–358.

Agostinelli, E., G. Tempera, N. Viceconte et al. 2010. Potential anticancer application of polyamine oxidation products formed by amino oxidase: A new therapeutic approach. *Amino Acids* 38: 353–368.

Agostinelli, E., F. Vianello, G. Magliulo, T. Thomas, and T. J. Thomas. 2015. Nanoparticle strategies for cancer therapeutics: Nucleic acids, polyamines, bovine serum oxidase and iron. *Int J Oncol* 46: 5–16.

Agrawal, N., P. V. N. Dasaradhi, A. Mohmmed, P. Malhotra, R. K. Bhatnagar, and S. K. Mukherjee. 2003. RNA interference: Biology, mechanism, and applications. *Microbiol Mol Biol Rev* 67: 657–685.

Agudelo-Romero, P., C. Bortolotti, M. S. Pais, A. F. Tiburcio, and A. M. Fortes. 2013. Study of polyamines during grape ripening indicate an important role of polyamine catabolism. *Plan Physiol Biochem* 67: 105–119.

Ahmadabadi, A., S. Ruf, and R. Bock. 2007. A leaf based regeneration and transformation system for maize (*Zea mays* L.). *Transgenic Res* 16: 437–448.

Ahn, I. P., S. Kim, W. B. Choi, and Y. H. Lee. 2003. Calcium restores prepenetration morphogenesis abolished by polyamines in *Colletotrichum gloeosporioides* infecting red pepper. *FEMS Microbiol Lett* 227: 237–241.

Ahn, H. J., K. H. Kim, J. Lee et al. 2004. Crystal structure of agmatinase reveals structural conservation and inhibition mechanism of the ureohydrolase superfamily. *J Biol Chem* 279: 50505–50513.

Ahuja, I., R. C. H. de Vos, A. M. Bones, and R. D. Hall. 2010. Plant molecular stress responses face climate change. *Plant Sci* 15: 664–674.

Aikens, D., S. Bunce, F. Onasch, R. Parker, 3rd, C. Hurwitz, and S. Clemans. 1983. The interactions between nucleic acids and polyamines. II. Protonation constants and 13C-NMR chemical shift assignments of spermidine, spermine, and homologs. *Biophys Chem* 17: 67–74.

Alabadi, D. and J. Carbonell. 1999. Differential expression of two spermidine synthase genes during early fruit development and in vegetative tissues of pea. *Plant Mol Biol* 39: 933–943.

Albeck, S., O. Dym, T. Unger, Z. Snapir, Z. Bercovich, and C. Kahana. 2008. Crystallographic and biochemical studies revealing the structural basis for antizyme inhibitor function. *Protein Sci* 17: 793–802.

Alcazar, R., F. Marco, J. C. Cuevas et al. 2006. Involvement of polyamines in plant response to abiotic stress. *Biotechnol Lett* 28: 1867–1876.

Alcazar, R., T. Altabella, F. Marco et al. 2010. Polyamines: Molecules with regulatory functions in plant abiotic stress tolerance. *Planta* 231: 1237–1249.

Aldesuquy, H., S. Haroun, S. Abo-Hamed, and A.-W. El-Saied. 2014. Involvement of spermine and spermidine in the control of productivity and biochemical aspects of yielded grains of wheat plants irrigated waste water. *EJBAS* I: 16–28.

Alet, A. I., D. H. Sánchez, J. C. Cuevas et al. 2012. New insights into the role of spermine in *Arabidopsis thaliana* under long-term salt stress. *Plant Sci* 182: 94–100.

Alvarez, F. J., L. M. Douglas, and J. B. Konopka. 2007. Sterol-rich plasma membrane domains in fungi. *Eukaryot Cell* 6: 755–763.

Amoroso, M. J. and C. M. Abate. 2012. Bioremediation of copper, chromium and cadmium by actinomycetes from contaminated soils. In *Bio-Geo Interactions in Metal-Contaminated Soils*, eds. E. Kothe and A. Varma, 349–364. Berlin: Springer Berlin Heidelberg.

An, Z. F., C. Y. Li, L. X. Zhang, and A. K. Alva. 2012. Role of polyamines and phospholipase D in maize (*Zea mays* L.) response to drought stress. *S Afr J Bot* 83: 145–150.

Andersen, S. M., R. Taylor, E. Holen, A. Aksnes, and M. Espe. 2014. Arginine supplementation and exposure time effects polyamine and glucose metabolism in primary liver cells isolated from Atlantic salmon. *Amino Acids* 46: 1225–1233.

Andréll, J., M. G. Hicks, T. Palmer, E. P. Carpenter, S. Iwata, and M. J. Maher. 2009. Crystal structure of the acid-induced arginine decarboxylase from *Escherichia coli*: Reversible decamer assembly controls enzyme activity. *Biochemistry* 48: 3915–3927.

Anehus, S., H. Emanuelsson, L. Persson, F. Sundler, I. E. Scheffler, and O. Heby. 1984. Localization of ornithine decarboxylase in mutant CHO cells that overproduce the enzyme. Differences between the intracellular distribution of monospecific ornithine decarboxylase antibodies and radiolabeled alpha-difluoromethylornithine. *Eur J Cell Biol* 35: 264–272.

Angelini, R., R. Federico, and P. Bonfante. 1995. Maize polyamine oxidase-antibody-production and ultrasound localization. *J Plant Physiol* 145: 686–692.

Angus-Hill, M. L., R. N. Dutnall, S. T. Tafrov, R. Sternglanz, and V. Ramakrishnan. 1999. Crystal structure of the histone acetyltransferase Hpa2: A tetrameric member of the Gcn5-related *N*-acetyltransferase superfamily. *J Mol Biol* 294: 1311–1325.

Antequera, F., M. Tamame, J. R. Villanueva, and T. Santos. 1984. DNA methylation in the fungi. *J Biol Chem* 259: 8033–8036.

Antony, T., W. Hoyer, D. Cherny, G. Heim, T. M. Jovin, and V. Subramaniam. 2003. Cellular polyamines promote the aggregation of alpha-synuclein. *J Biol Chem* 278: 3235–3240.

Aoki, F. Y. and D. S. Sitar. 1988. Clinical pharmacokinetics of amantadine hydrochloride. *Clin Pharmacokinet* 14: 35–51.

Aouida, M., A. Leduc, R. Poulin, and D. Ramotar. 2005. AGP2 encodes the major permease for high affinity polyamine import in *Saccharomyces cerevisiae*. *J Biol Chem* 280: 24267–24276.

Auvinen, M., A. Paasinen, L. C. Andersson, and E. Holtta. 1992. Ornithine decarboxylase activity is critical for cell transformation. *Nature* 360: 355–358.

Bacchi, C. J., J. Garofalo, D. Mockenhaupt et al. 1983. *In vivo* effects of alpha-DL-difluoromethylornithine on the metabolism and morphology of *Trypanosoma brucei*. *Mol Biochem Parasitol* 7: 209–225.

Bachrach, U. 2005. Naturally occurring polyamines: Interaction with macromolecules. *Curr Protein Pept Sci* 6: 559–566.

Bachrach, U. 2010. The early history of polyamine research. *Plant Physiol Biochem* 48: 490–495.

Bachrach, U., Y. C. Wang, and A. Tabib. 2001. Polyamines: New cues in cellular signal transduction. *News Physiol Sci* 16: 106–109.

Backlund, P. S., Jr. and R. A. Smith. 1981. Methionine synthesis from 5′-methylthioadenosine in rat liver. *J Biol Chem* 256: 1533–1535.

Bahn, Y. S., C. Xue, A. Idnurm, J. C. Rutherford, J. Heitman, and M. E. Cardenas. 2007. Sensing the environment: Lessons from fungi. *Nat Rev Microbiol* 5: 57–69.

Bailey, A., E. Mueller, and P. Bowyer. 2000. Ornithine decarboxylase of *Stagonospora (Septoria) nodorum* is required for virulence toward wheat. *J Biol Chem* 275: 14242–14247.

Balasundaram, D. and A. K. Tyagi. 1991. Polyamine–DNA nexus: Structural ramifications and biological implications. *Mol Cell Biochem* 100: 129–140.

Balasundaram, D., C. W. Tabor, and H. Tabor. 1991. Spermidine or spermine is essential for the aerobic growth of *Saccharomyces cerevisiae*. *Proc Natl Acad Sci U S A* 88: 5872–5876.

Balasundaram, D., C. W. Tabor, and H. Tabor. 1993. Oxygen-toxicity in a polyamine-depleted Spe2-delta mutant of *Saccharomyces-cerevisiae*. *Proc Natl Acad Sci U S A* 90: 4693–4697.

Balasundaram, D., C. W. Tabor, and H. Tabor. 1996. Sensitivity of polyamine-deficient *Saccharomyces cerevisiae* to elevated temperatures. *J Bacteriol* 178: 2721–2724.

Bale, S., M. M. Lopez, G. I. Makhatadze, Q. Fang, A. E. Pegg, and S. E. Ealick. 2008. Structural basis for putrescine activation of human *S*-adenosylmethionine decarboxylase. *Biochemistry* 16: 13404–13417.

Barnett, G. R. and M. N. Kazarinoff. 1984. Purification and properties of ornithine decarboxylase from *Physarum polycephalum*. *J Biol Chem* 259: 179–183.

Bartholeyns, J. 1983. Treatment of metastatic Lewis lung carcinoma with DL-alfa-difluoromethylornithine. *Eur J Cancer* 19: 567–572.

Bastola, D. R. and S. C. Minocha. 1995. Increase putrescine biosynthesis through transfer of mouse ornithine decarboxylase cDNA in carrot provides tolerance embryogenesis. *Plant Physiol* 109: 63–71.

Basu, H. S., H. C. Schwietert, B. G. Feuerstein, and L. J. Marton. 1990. Effects of variation in the structure of spermine on the association with DNA and the induction of DNA conformational changes. *Biochem J* 269: 329–334.

Basu, H. S., I. V. Smirnov, H. F. Peng, K. Tiffany, and V. Jackson. 1997. Effects of spermine and its cytotoxic analogs on nucleosome formation on topologically stressed DNA in vitro. *Eur J Biochem* 243: 247–258.

Bauer, P. M., G. M. Buga, and L. J. Ignarro. 2001. Role of p42/p44 mitogen activated protein kinase and p21waf1/cip1 in the regulation of vascular smooth muscle cell proliferation by nitric oxide. *Proc Natl Acad Sci U S A* 98: 12802–12807.

Baugh, L., I. Phan, D. W. Begley et al. 2014. Increasing the structural coverage of tuberculosis drug targets. *Tuberculosis (Edinb)* 95: 142–148.

Belda-Palazon, B., L. Ruiz, E. Marti et al. 2012. Aminopropyltransferases involved in polyamine biosynthesis localize preferentially in the nucleus of plant cells. *PLoS One* 7: e46907.

Bell, M. R., J. A. Belarde, H. F. Jhonson, and C. D. Aizenman. 2011. A neuroprotective role of polyamines in a *Xenopus* tadpole model of epilepsy. *Nat Neurosci* 14: 505–512.

Belotserkovskaya, R., D. E. Sterner, M. Deng, M. H. Sayre, P. M. Lieberman, and S. L. Berger. 2000. Inhibition of TATA-binding protein function by SAGA subunits Spt3 and Spt8 at Gcn-4 activated promoters. *Mol Cell Biol* 20: 634–647.

Bencini, A., A. Bianchi, E. Garcia-Espana, M. Micheloni, and J. A. Ramirez. 1999. Proton coordination by polyamine compounds in aqueous solution. *Coord Chem Rev* 188: 97–156.

Beninati, S., A. Abbruzzese, and M. Cardinali. 1993. Differences in the post-translational modification of proteins by polyamines between weakly and highly metastatic B16 melanoma cells. *Int J Cancer* 53: 792–797.

Bennett, J. W. 1983. *Secondary Metabolism and Differentiation in Fungi*. Boca Raton, FL: CRC Press/Taylor & Francis.

Berger, S. L. 2007. The complex language of chromatin regulation during transcription. *Nature* 447: 407–412.

Berman, H. M., J. Westbrook, Z. Feng et al. 2000. The Protein Data Bank. *Nucleic Acids Res* 28: 235–242.

Bernstein, H. G. and M. Muller. 1999. The cellular localization of the L-ornithine decarboxylase/polyamine system in normal and diseased central nervous systems. *Prog Neurobiol* 57: 485–505.

Bernstein, E., A. A. Caudy, S. M. Hammond, and G. J. Hannon. 2001. Role for a bidentate ribonuclease in the initiation step of RNA interference. *Nature* 409: 363–366.

Berntsson, P. S. H., K. Alm, and S. M. Oredsson. 1999. Half-lives of ornithine decarboxylase and *S*-adenosylmethionine decarboxylase activities during the cell cycle of Chinese hamster ovary cells. *Biochem Biophys Res Commun* 263: 13–16.

Bertucci Barbosa, L. C., L. Alves Dos Santos, J. De Souza Ferreira, K. Fernandes Viana, A. S. Rodrigues Cangussu, and R. Wagner de Souza. 2014. A web-based resource for structural information on eIF5A and its related proteins: New potential therapeutic targets in many human disorders. *Int J Pharm Pharm Sci* 6: 610–612.

Berwanger, A., S. Eyrisch, I. Schuster, V. Helms, and R. Bernhardt. 2010. Polyamines: Naturally occurring small molecule modulators of electrostatic protein–protein interactions. *J Inorg Biochem* 104: 118–125.

Bettuzzi, S., M. Marinelli, P. Strocchi, P. Davalli, D. Cevolani, and A. Corti. 1995. Different localization of spermidine/spermine N^1-acetyltransferase and ornithine decarboxylase transcripts in the rat kidney. *FEBS Lett* 377: 321–324.

Bettuzzi, P. S., P. Davalli, S. Astancolle et al. 1999. Coordinate changes of polyamine metabolism regulatory proteins during the cell cycle of normal human dermal fibroblast. *FEBS Lett* 446: 18–22.

Bewley, M. C., V. Graziano, J. S. Jiang et al. 2006. Structures of wild type and mutant human spermidine/spermine N^1-acetyltransferase, a potential therapeutic drug target. *Proc Natl Acad Sci U S A* 103: 2063–2068.

Bhart, A., R. K. Kaul, M. Reshi, and N. Gupta. 2014. Effect of polyamines on shelf life and chilling injury of mango cv dashehari. *Bioscan* 9: 1097–1100.

Bharti, and M. V. Rajam. 1995. Effects of the polyamine biosynthesis inhibitor difluoromethylornithine on growth, polyamine levels, chromosome behavior and polygenic traits in wheat (*Triticuma estivum* L.). *Ann Bot* 76: 297–301.

Bhattacharya, E. and M. V. Rajam. 2007. Polyamine biosynthetic pathway: A potential target for enhancing alkaloid production. In *Application of Plant Metabolic Engineering*, eds. R. Verpoorte, A. W. Alfermann, and T. S. Johnson, 129–143. Netherlands: Springer.

Bianchi, M., R. Amendola, R. Federico, F. Polticelli, and P. Mariottini. 2005. Two short protein domains are responsible for the nuclear localization of the mouse spermine oxidase mu isoform. *FEBS J* 272: 3052–3059.

Bianchi, M., F. Polticelli, P. Ascenzi, M. Botta, R. Federico, P. Mariottini, and A. Cona. 2006. Inhibition of polyamine and spermine oxidases by polyamine analogues. *FEBS J* 273: 1115–1123.

Biasi, R., G. Costa, and N. Bagni. 1991. Polyamine metabolism as related to fruit set and growth. *Plant Physiol Biochem* 19: 497–506.

Bibi, A. C., D. M. Oosterhuis, and E. D. Gonias. 2010. Exogenous application of putrescine ameliorates the effect of high temperature in *Gossypium hirsutum* L. Flowers and fruit development. *J Agron Crop Sci* 196: 205–211.

Biondi, S., I. Polgrossi, and N. Bagni. 1993. Effect of polyamine biosynthesis inhibitors on mycelial growth and concentrations of polyamines in *Ophiostoma ulmi*. *New Phytol* 123: 415–419.

Birecka, H., M. O. Garraway, R. J. Baumann, and P. P. McCann. 1986. Inhibition of ornithine decarboxylase and growth of the fungus *Helminthosporium maydis*. *Plant Physiol* 80: 798–800.

Birkholtz, L. M., C. Wrenger, F. Joubert, G. A. Wells, R. D. Walter, and A. I. Louw. 2004. Parasite-specific inserts in the bifuntional *S*-adenosylmethionine decarboxylase/ornithine decarboxylase of *Plasmodium falciparum* modulate catalytic activities and domain interactions. *Biochem J* 15: 439–448.

Blagbrough, I. S. and A. J. Geall. 2003. Acidity of basic polyamines: pKa studies of symmetrical and unsymmetrical polyamines and their conjugates compared with their DNA binding. *Abstr Pap Am Chem S* 225: U199.

Blagbrough, I. S., A. A. Metwally, and A. J. Geall. 2011. Measurement of polyamine pKa values. *Methods Mol Biol* 720: 493–503.

Bolkenius, F. N. and N. Seiler. 1981. Acetylderivatives as intermediates in polyamine catabolism. *Int J Biochem* 13: 287–292.

Borrell, A., F. A. Culiañez-Macia, T. Altabella, R. T. Besford, D. Flores, and A. F. Tiburcio. 1995. Arginine decarboxylase is localized in chloroplasts. *Plant Physiol* 109: 771–776.

Bosch, J. and W. G. J. Hol. Structural Genomics of Pathogenic Protozoa Consortium. 2004. Structural analysis of *Leishmania mexicana* eukaryotic initiation factor 5A. RCSB Protein Data Bank. Available at http://www.rcsb.org/.

Bose, I. and T. L. Doering. 2011. Efficient implementation of RNA interference in the pathogenic yeast *Cryptococcus neoformans*. *J Microbiol Methods* 86: 156–159.

Bottiglieri, T. 2002. *S*-adenosyl-L-methionine (SAMe): From the bench to the bebside—Molecular basis of a pleiotrophic molecule. *Am J Clin Nutr* 76: 1151S–1157S.

Brachet, P., H. Debbabi, and D. Tome. 1995. Transport and steady-state accumulation of putrescine in brush-border membrane vesicles of rabbit small intestite. *Am J Physiol* 269: G754–G762.

Bras, A. P., H. R. Hoff, F. Y. Aoki, and D. S. Sitar. 1998. Amantadine acetylation may be effected by acetyltransferases other than NAT1 or NAT2. *Can J Physiol Pharmacol* 76: 701–706.

Bras, A. P., J. Janne, C. W. Porter, and D. S. Sitar. 2001. Spermidine/spermine *N*(1)-acetyltransferase catalyzes amantadine acetylation. *Drug Metab Dispos* 29: 676–680.

Braunlin, W. H., T. J. Strick, and M. T. Record, Jr. 1982. Equilibrium dialysis studies of polyamine binding to DNA. *Biopolymers* 21: 1301–1314.

Braus, G. H., S. Irniger, and O. Bayram. 2010. Fungal development and the COP9 signalosome. *Curr Opin Microbiol* 13: 672–676.

Bregoli, A. M., S. Scaramagli, G. Costa, E. Sabatini, V. Ziosi, S. Biondi, and P. Torrigiani. 2002. Peach (*Prunus persica* L.) fruit ripening: Aminoethoxyvinylglycine (AVG) and exogenous polyamines affect ethylene emission and flesh firmness. *Physiol Plant* 114: 472–481.

Brieger L. 1885. *Weitere Untersuchungen über Ptomaine* [*Further Investigations into Ptomaines*]. Berlin: August Hirschwald.

Brogaard, K. R., L. Xi, J. P. Wang, and J. Widom. 2012. A chemical approach to mapping nucleosomes at base pair resolution in yeast. *Methods Enzymol* 513: 315–334.

Brooks, B. 1994. A model for systemic lupus erythematosus based on chromatin disruption by polyamines. *Med Hypotheses* 43: 403–406.

Bueb, J. L., M. Mousli, and Y. Landry. 1991. Molecular basis for cellular effects of naturally occurring polyamines. *Agents Actions* 33: 84–87.

Burger, P. B., L. M. Birkholtz, F. Joubert, N. Haider, R. D. Walter, and A. I. Louw. 2007. Structural and mechanistic insights into the action of *Plasmodium faciparum* spermidine synthase. *Bioorg Med Chem* 15: 1628–1637.

Burton, D. R., S. Forsen, and P. Reimarsson. 1981. The interaction of polyamines with DNA: A 23Na NMR study. *Nucleic Acids Res* 9: 1219–1228.

Busch, S., S. E. Eckert, S. Krappmann, and G. H. Braus. 2003. The COP9 signalosome is an essential regulator of development in the filamentous fungus *Aspergillus nidulans*. *Mol Microbiol* 49: 717–730.

Cady, S. D., K. Schmidt-Rohr, J. Wang, C. S. Soto, F. DeGrado, and M. Hong. 2009. Structure of the amantadine binding site of influenza M2 proton channels in lipid bilayers. *Nature* 463: 689–692.

Calcabrini, A., G. Arancia, M. Marra, P. Crateri, O. Befani, A. Martone, and E. Agostinelli. 2002. Enzymatic oxidation products of spermine induce greater cytotoxic effects on human multidrug-resistant colon carcinoma cells (LoVo) than on their wild-type counterparts. *Int J Cancer* 93: 43–45.

Calvo-Mendez, C. and J. Ruiz-Herrera. 1991. Regulation of *S*-adenosylmethionine decarboxylase during the germination of sporangiospores of *Mucor rouxii*. *J Gen Microbiol* 137: 307–314.

Calvo-Mendez, C., M. Martinez-Pacheco, and J. Ruiz-Herrera. 1987. Regulation of ornithine decarboxylase activity in *Mucor bacilliformis* and *Mucor rouxii*. *Exp Mycol* 11: 270–277.

Cano, C., L. Herrera-Estrella, and J. Ruiz-Herrera. 1988. DNA methylation and polyamines in regulation of development of the fungus *Mucor rouxii*. *J Bacteriol* 170: 5946–5948.

Cano-Canchola, C., L. Sosa, W. Fonzi, P. Sypherd, and J. Ruiz-Herrera. 1992. Developmental regulation of *CUP* gene expression through DNA methylation in *Mucor spp*. *J Bacteriol* 174: 362–366.

Caraglia, M., M. Marra, G. Giuberti et al. 2003. The eukaryotic initiation factor 5A is involved in the regulation of proliferation and apoptosis induced by interferon and EGF in human cancer cells. *J Biochem* 133: 757–765.

Caraglia, M., M. H. Park, E. C. Wolff, M. Marra, and A. Abbruzzese. 2013. eIF5A isoforms and cancer: Two brothers for two functions? *Amino Acids* 44: 103–109.

Carbonell, J. and J. Navarro. 1989. Correlation of spermine levels with ovary senescence and with fruit set and development in *Pisum sativum* L. *Planta* 178: 482–487.

Casero, R. A., Jr. and A. E. Pegg. 1993. Spermidine/spermine N^1-acetyltransferase—The turning point in polyamine metabolism. *FASEB J* 7: 653–661.

Casero, R. A. and L. J. Marton. 2007. Targeting polyamine metabolism and function in cancer and other hyperproliferative diseases. *Nat Rev Drug Discov* 6: 373–390.

Casero, R. A., Y. Wang, T. M. Steward et al. 2003. The role of polyamine catabolism in antitumor drug response. *Biochem Soc Trans* 31: 361–365.

Castelli, A. and C. Rossoni. 1968. Polyamines and nucleic acids in the growing yeast. *Experientia* 24: 1119–1120.

Cedar, H. and Y. Bergman. 2009. Linking DNA methylation and histone modification: Patterns and paradigms. *Nat Rev Genet* 10: 295–304.

Celano, P., S. B. Baylin, and R. A. Casero, Jr. 1989. Polyamines differentially modulate the transcription of growth-associated genes in human colon carcinoma cells. *J Biol Chem* 264: 8922–8927.

Celano, P., C. M. Berchtold, F. M. Giardiello, and R. A. Casero, Jr. 1989. Modulation of growth gene expression by selective alteration of polyamines in human colon carcinoma cells. *Biochem Biophys Res Commun* 165: 384–390.

Cerrada-Gimenez, M., M. Pietila, S. Loimas et al. 2011. Continuous oxidative stress due to activation of polyamine catabolism accelerates aging and protects against hepatotoxic insults. *Transgenic Res* 20: 387–396.

Cervelli, M., A. Bellini, M. Bianchi, L. Marcocci, S. Nocera, F. Polticelli, R. Federico, R. Amendola, and P. Mariottini. 2004. Mouse spermine oxidase gene splice variants. Nuclear subcellular localization of a novel active isoform. *Eur J Biochem* 271: 760–770.

Cervelli, M., D. Salvi, F. Polticelli, R. Amendola, and P. Mariottini. 2013. Structure–function relationships in the evolutionary framework of spermine oxidase. *J Mol Biol* 76: 365–370.

Cetrullo, S., A. Facchini, I. Stanic, B. Tantini, C. Pignatti, C. M. Caldarera, and F. Flamigni. 2010. Difluoromethylornithine inhibits hypertrophic, pro-fibrotic and pro-apoptotic actions of aldosterone in cardiac cells. *Amino Acids* 38: 525–531.

Cetrullo, S., B. Tantini, A. Facchini, C. Pignatti, C. Stefanelli, C. M. Caldarera, and F. Flamingni. 2011. A pro-survival effect of polyamine depletion on norepinephrine-mediated apoptosis in cardiac cells: Role of signaling enzymes. *Amino Acids* 40: 1127–1137.

Cha, H. J., J. Jeong, C. Rojviriya, and Y. Kim. 2014. Structure of putrescine aminotransferase from *Escherichia coli* provides insights into the substrate specificity among Class III aminotransferases. *PLoS One* 9: e113212.

Chadwick, J., M. Jones, A. E. Mercer, P. A. Stocks, S. A. Ward, B. K. Park, and P. M. O'Neill. 2010. Design synthesis and antimalarial/anticancer evaluation of spermidine linked artemisinin conjugates designed to exploit polyamine transporters in *Plasmodium falciparum* and HL-60 cancer cell lines. *Bioorg Med Chem* 18: 2586–2597.

Chang, Z.-F. and K. Y. Chen. 1988. Regulation of ornithine decarboxylase and other cell cycle-dependent genes during senescence of IMR-90 human diploid fibroblasts. *J Biol Chem* 263: 11431–11435.

Chang, S. S., Z. Zhang, and Y. Liu. 2012. RNA interference pathways in fungi: Mechanisms and functions. *Annu Rev Microbiol* 66: 305–323.

Chao, W. S., M. E. Foley, D. P. Horvath, and J. V. Anderson. 2007. Signals regulating dormancy in vegetative buds. *Int J Plant Dev Biol* 1: 49–56.

Chattopadhyay, M. K., C. W. Tabor, and H. Tabor. 2002. Absolute requirement of spermidine for growth and cell cycle progression of fission yeast (*Schizosaccharomyces pombe*). *Proc Natl Acad Sci U S A* 99: 10330–10334.

Chattopadhyay, M. K., C. W. Tabor, and H. Tabor. 2003a. Spermidine but not spermine is essential for hypusine biosynthesis and growth in *Saccharomyces cerevisiae*: Spermine is converted to spermidine *in vivo* by the FMS1-amine oxidase. *Proc Natl Acad Sci U S A* 100: 13869–13874.

Chattopadhyay, M. K., C. W. Tabor, and H. Tabor. 2003b. Polyamines protect *Escherichia coli* cells from the toxic effect of oxygen. *Proc Natl Acad Sci U S A* 100: 2261–2265.

Chattopadhyay, M. K., C. W. Tabor, and H. Tabor. 2006a. Polyamine deficiency leads to accumulation of reactive oxygen species in a spe2 Delta mutant of *Saccharomyces cerevisiae*. *Yeast* 23: 751–761.

Chattopadhyay, M. K., C. W. Tabor, and H. Tabor. 2006b. Methylthioadenosine and polyamine biosynthesis in a *Saccharomyces cerevisiae meu1*delta mutant. *Biochem Biophys Res Commun* 343: 203–207.

Chattopadhyay, M. K., M. H. Park, and H. Tabor. 2008. Hypusine modification for growth is the major function of spermidine in *Saccharomyces cerevisiae* polyamine auxotrophs grown in limiting spermidine. *Proc Natl Acad Sci U S A* 105: 6554–6559.

Chattopadhyay, M. K., W. Chen, G. Poy, M. Cam, D. Stiles, and H. Tabor. 2009. Microarray studies on the genes responsive to the addition of spermidine or spermine to a *Saccharomyces cerevisiae* spermidine synthase mutant. *Yeast* 26: 531–544.

Chattopadhyay, M. K., C. Fernandez, D. Sharma, P. McPhie, and H. Tabor. 2011a. Antizyme, an unusual regulatory protein: Studies on the yeast antizyme–ornithine decarboxylase complex. *FASEB J* 25: 754.2.

Chattopadhyay, M. K., C. Fernandez, D. Sharma, P. McPhie, and D. C. Masison. 2011b. Yeast ornithine decarboxylase and antizyme form 1:1 complex in vitro: Purification and characterization of the inhibitory complex. *Biochem Biophys Res Commun* 406: 177–182.

Chen, Y., D. L. Kramer, J. Jell, S. Vujcic, and C. W. Porter. 2003. Small interfering RNA suppression of polyamine analog-induced spermidine/spermineN^1-acetyltransferase. *Mol Pharmacol* 64: 1153–1159.

Cheng, L., Y. Zou, S. Ding, J. Zhang, X. Yu, J. Cao, and G. Lu. 2009. Polyamine accumulation in transgenic tomato enhances the tolerance to high temperature stress. *J Integr Plant Biol* 51: 489–499.

Chiang, P. K., R. K. Gordon, J. Tal, G. C. Zeng, B. P. Doctor, K. Pardhasaradhi, and P. P. McCann. 1996. *S*-Adenosylmethionine and methylation. *FASEB J* 10: 471–480.

Chibucos, M. C. and P. F. Morris. 2006. Levels of polyamines and kinetic characterization of their uptake in the soybean pathogen *Phytophthora sojae*. *Appl Environ Microbiol* 72: 3350–3356.

Childs, A. C., D. J. Mehta, and E. W. Gerner. 2003. Polyamine-dependent gene expression. *Cell Mol Life Sci* 60: 1394–1406.

Chiu, S. and N. L. Oleinick. 1998. Radioprotection of cellular chromatin by the polyamines spermine and putrescine: Preferential action against formation of DNA–protein cross-links. *Radiat Res* 149: 543–549.

Choi, W. B., S. H. Kang, Y. W. Lee, and Y. H. Lee. 1998. Cyclic AMP restores appressorium formation inhibited by polyamines in *Magnaporthe grisea*. *Phytopathology* 88: 58–62.

Choudhary, S. P., M. Kanwar, R. Bhardwaj, J.-Q. Yu, and L.-S. Phan Tran. 2012. Chromium stress mitigation by polyamine-brassinosteroid application involves phytohormonal and physiological strategies in *Raphanus sativus* L. *PLoS One* 7: 1–11.

Chowhan, R. K. and L. R. Singh. 2013. Polyamines in modulating protein aggregation. *J Proteins Proteom* 3: 141–150.

Clark, G. R., D. G. Brown, M. R. Sanderson et al. 1990. Crystal and solution structures of the oligonucleotide d(ATGCGCAT)2: A combined X-ray and NMR study. *Nucleic Acids Res* 18: 5521–5528.

Clay, N. K. and T. Nelson. 2014. *Arabidopsis* thick vein mutation affects vein thickness and organ vascularization, and resides in a provascular cell-specific spermine synthase involved in vein definition and in polar Auxin Transport. *Plant Physiol* 138: 767–777.

Clemens, J. C., C. A. Worley, N. S. Leff, M. Muda, T. Maehama, B. A. Hemmings, and J. E. Dixon. 2000. Use of double-stranded RNA interference in Drosophila cell lines to dissect signal transduction pathways. *Proc Natl Acad Sci U S A* 97: 6499–6503.

Coffino, P. 2001a. Antizyme, a mediator of ubiquitin-independent proteasomal degradation. *Biochimie* 83: 319–323.

Coffino, P. 2001b. Regulation of cellular polyamines by antizyme. *Nat Rev Mol Cell Biol* 2: 188–194.

Cole, G. T. 1986. Models of cell differentiation in conidial fungi. *Microbiol Rev* 50: 95–132.

Coleman, C. S., B. A. Stanley, R. Viswanath, and A. E. Pegg. 1994. Rapid exchange of subunits of mammalian ornithine decarboxylase. *J Biol Chem* 269: 3155–3158.

Coleman, C. S., B. A. Stanley, A. D. Jones, and A. E. Pegg. 2004. Spermidine/spermine-*N*[1]-acetyltransferase-2 (SSAT2) acetylates thialysine and is not involved in polyamine metabolism. *Biochem J* 384: 139–148.

Colombo, R., R. Cerana, and N. Bagni. 1992. Evidence for polyamine channels in protoplasts and vacuoles of *Arabidopsis thaliana* cells. *Biochem Biophys Res Commun* 182: 1187–1192.

Colton, C. A., Q. Xu, J. R. Burke et al. 2004. Disrupted spermine homeostasis: A novel mechanism in polyglutamine-mediated aggregation and cell death. *J Neurosci* 24: 7118–7127.

Cona, A., F. Manetti, R. Leone, F. Corelli, P. Tavladoraki, F. Polticelli, and M. Botta. 2004. Molecular basis for the binding of competitive inhibitors of maize polyamine oxidase. *Biochemistry* 43: 3426–3435.

Cone, M. C., K. Marchitto, B. Zehfus, and A. J. Ferro. 1982. Utilization by *Saccharomyces cerevisiae* of 5′-methylthioadenosine as a source of both purine and methionine. *J Bacteriol* 151: 510–515.

Cox, D. R., T. Trouillot, P. L. Ashley, M. Brabant, and P. Coffino. 1988. A functional mouse ornithine decarboxylase gene (*Odc*) maps to chromosome 12: Further evidence of homology between mouse chromosome 12 and the short arm of human chromosome 2. *Cytogenet Cell Genet* 48: 92–94.

Cu, C., R. Bähring, and M. L. Mayer. 1998. The role of hydrophobic interactions in binding of polyamines to non NMDA receptor ion channels. *Neuropharmacology* 37: 1381–1391.

Cuevas, J. C., R. Lopez-Cobollo, R. Alcazar et al. 2008. Putrescine is involved in *Arabidopsis* freezing tolerance and cold acclimation by regulating abscisic acid levels in response to low temperature. *Plant Physiol* 148: 1094–1105.

Cullen, B. R. 2002. RNA interference: Antiviral defense and genetic tool. *Nat Immunol* 3: 597–599.

Cullis, P. M., R. E. Green, L. Merson-Davies, and N. Travis. 1999. Probing the mechanism of transport and compartmentalisation of polyamines in mammalian cells. *Chem Biol* 6: 717–729.

Culver, G. M. 2003. Assembly of the 30S ribosomal subunit. *Biopolymers* 68: 234–249.

D'Agostino, L. and A. Di Luccia. 2002. Polyamines interact with DNA as molecular aggregates. *Eur J Biochem* 269: 4317–4325.

D'Agostino, L., M. di Pietro, and A. Di Luccia. 2006. Nuclear aggregates of polyamines. *IUBMB Life* 58: 75–82.

D'Antonio, E. L., B. Ullman, S. C. Roberts, U. G. Dixit, M. E. Wilson, Y. Hai, and D. W. Christianson. 2013. Crystal structure of arginase from *Leishmania mexicana* and implications for the inhibition of polyamine biosynthesis in parasitic infections. *Arch Biochem Biophys* 535: 163–176.

Da'Dara, A. A. and R. D. Walter. 1998. Molecular and biochemical characterization of *S*-adenosylmethionine decarboxylase from the free-living nematode *Caenorhabditis elegans*. *Biochem J* 336: 545–550.

Dailey, T. A. and H. A. Dailey. 1998. Identification of an FAD superfamily containing protoporphyrinogen oxidases, monoamine oxidases, and phytoene desaturase. Expression and characterization of phytoene desaturase of *Myxococcus xanthus*. *J Biol Chem* 273: 13658–13662.

Das, U., G. Hariprasad, A. S. Ethayathulla et al. 2007. Inhibition of protein aggregation: Supra molecular assemblies of arginine hold the key. *PLoS One* 2007: 1–9.

Davary Nejad, G., M. Zarei, E. Ardakani, and M. E. Nasrabadi. 2013. Influence of putrescine application on storability, postharvest quality and antioxidant activity of two apricot (*Prunus armeniaca* L.) cultivars. *Not Sci Biol* 5: 212–219.

Davis, R. H. 1986. Compartmental and regulatory mechanisms in the arginine pathways of *Neurospora crassa* and *Saccharomyces cerevisiae*. *Microbiol Rev* 50: 280–313.

Davis, R. H., M. B. Lawless, and L. A. Port. 1970. Arginaseless *Neurospora*: Genetics, physiology and polyamine synthesis. *J Bacteriol* 102: 299–305.

de Alencastro, G., D. E. McCloskey, S. E. Kliemann et al. 2008. New SMS mutation leads to striking reduction in spermine synthase protein function and a severe form of Snyder-Robinson X-linked recessive mental retardation syndrome. *J Med Genet* 45: 539–543.

DeLano, W. L. 2002. *The PyMOL Molecular Graphics System*. San Carlos, CA: DeLano Scientific LLC. Available at http://www.pymol.org Open-Source PyMOL 0.99rc6. PyMOL(TM).

Demey, F. 1987. Studies on the efficacy of DL-alpha-difluoromethylornithine (DFMO) associated with bleomycin and suram in for treatment of mice infected with metacyclic forms of *Trypanosoma brucei*. *Vet Res Commun* 11: 217–214.

Deng, H., V. A. Bloomfield, J. M. Benevides, and G. J. Thomas, Jr. 2000. Structural basis of polyamine–DNA recognition: Spermidine and spermine interactions with genomic B-DNAs of different GC content probed by Raman spectroscopy. *Nucleic Acids Res* 28: 3379–3385.

Deryng, D., D. Conway, N. Ramankutty, J. Price, and R. Warren. 2014. Global crop yield response to extreme heat stress under multiple climate change futures. *Environ Res Lett* 9: 1–13.

DeScenzo, R. A. and S. C. Minocha. 1993. Modulation of cellular polyamines in tobacco by transfer and expression of mouse ornithine decarboxylase cDNA. *Plant Mol Biol* 22: 113–127.

Desiderio, M. A., P. Dansi, L. Tacchini, and A. Bernelli-Zazzera. 1999. Influence of polyamines on DNA binding of heat shock and activator protein 1 transcription factors induced by heat shock. *FEBS Lett* 45: 149–153.

Devens, B. H., R. S. Weeks, M. R. Burns, C. L. Carlson, and M. K. Brawer. 2000. Polyamine depletion therapy in prostate cancer. *Prostate Cancer Prostatic Dis* 3: 275–279.

DiGangi, J. J., M. Seyfzadeh, and R. H. Davis. 1987. Ornithine decarboxylase from *Neurospora crassa*. Purification, characterization, and regulation by inactivation. *J Biol Chem* 262: 7889–7893.

Do, P. T., O. Drechsel, A. G. Heyer, D. K. Hincha, and E. Zuther. 2014. Changes in free polyamine levels, expression of polyamine biosynthesis genes, and performance of rice cultivares under salt stress: A comparison with responses to drought. *Front Plant Sci* 5: 1–16.

Douki, T., Y. Bretonniere, and J. Cadet. 2000. Protection against radiation-induced degradation of DNA bases by polyamines. *Radiat Res* 153: 29–35.

Duan, J., J. Li, S. Guo, and Y. Kang. 2008. Exogenous spermidine affects polyamine metabolism in salinity-stressed *Cucumis sativus* roots and enhances short-term salinity tolerance. *J Plant Physiol* 165: 1620–1635.

Dudley, H. W., O. Rosenheim, and W. W. Starling. 1926. The chemical constitution of spermine. III. Structure and synthesis. *Biochem J* 20: 1082–1094.

Dudley, H. W., O. Rosenheim, and W. W. Starling. 1927. The constitution and synthesis of spermidine, a newly discovered base isolated from animal tissues. *Biochem J* 21: 97–103.

Dufe, V. T., K. Luersen, M. L. Eschbach, N. Haider, T. Karlberg, R. D. Walter, and S. Al-Karadagui. 2005. Cloning, expression, characterisation and three-dimensional structure determination of *Caenorhabditis elegans* spermidine synthase. *FEBS Lett* 579: 6037–6043.

Dufe, V. T., W. Qiu, I. B. Müller, R. Hui, R. D. Walter, and S. Al-Karadaghi. 2007. Crystal structure of *Plasmodium falciparum* spermidine synthase in complex with the substrate decarboxylated *S*-adenosylmethionine and the potent inhibitors 4MCHA and AdoDATO. *J Mol Biol* 373: 167–177.

Duguid, J. G. and V. A. Bloomfield. 1995. Aggregation of melted DNA by divalent metal ion-mediated cross-linking. *Biophys J* 69: 2642–2648.

Duguid, J., V. A. Bloomfield, J. Benevides, and G. J. Thomas, Jr. 1993. Raman spectroscopy of DNA–metal complexes. I. Interactions and conformational effects of the divalent cations: Mg, Ca, Sr, Ba, Mn, Co, Ni, Cu, Pd, and Cd. *Biophys J* 65: 1916–1928.

Duranton, B., G. Keith, Goss, C. Berqmann, R. Schleiffer, and F. Raul. 1998. Concomitant changes in polyamine pools and DNA methylation during growth inhibition of human colonic cancer cells. *Exp Cell Res* 243: 319–325.

Durie, B. G., S. E. Salmon, and D. H. Russell. 1977. Polyamines as markers of response and disease activity in cancer chemotherapy. *Cancer Res* 37: 214–221.

Echandi, G. and I. D. Algranati. 1975. Defective 30S ribosomal particles in a polyamine auxotroph of *Escherichia coli*. *Biochem Biophys Res Commun* 67: 1185–1191.

Ehinger, A., S. H. Denison, and G. S. May. 1990. Sequence, organization and expression of the core histone genes of *Aspergillus nidulans*. *Mol Gen Genet* 222: 416–424.

Eisenberg, T., H. Knauer, A. Schauer et al. 2009. Induction of autophagy by spermine promotes longevity. *Nat Cell Biol* 11: 1305–1314.

Ekstrom, J. L., I. I. Mathews, B. A. Stanley, A. E. Pegg, and S. E. Ealick. 1999. The crystal structure of human *S*-adenosylmethionine decarboxylase at 2.25 A resolution reveals a novel fold. *Structure* 7: 583–595.

El Ghachtouli, N., M. Paynot, J. Martin-Tanguy, D. Morandi, and S. Gianinazzi. 1996. Effect of polyamines and polyamine biosynthesis inhibitors on spore germination and hyphal growth of *Glomus mosseae*. *Mycol Res* 100: 597–600.

El Hadrami, I. and J. D'Auzac. 1992. Effects of polyamine biosynthetic inhibitors on somatic embryogenesis and cellular polyamines in *Hevea brasiliensis*. *J Plant Physiol* 140: 33–36.

El-Halfawy, O. M. and M. A. Valvano. 2014. Putrescine reduces antibiotic-induced oxidative stress as a mechanism of modulation of antibiotic resistance in *Burkholderia cenocepacia*. *Antimicrob Agents Chemother* 58: 4162–4171.

Elias, S., B. Bercovich, C. Kahana et al. 1995. Degradation of ornithine decarboxylase by the mammalian and yeast 26S proteasome complexes requires all the components of the protease. *Eur J Biochem* 229: 276–283.

El-Salahy, E. M. 2002. Correlation between polyamines and apoptosis among Egyptian breast cancer patients. *Clin Biochem* 35: 555–560.

Erez, O. and C. Kahana. 2001. Screening for modulators of spermine tolerance identifies Sky1, the SR protein kinase of *Saccharomyces cerevisiae*, as a regulator of polyamine transport and ion homeostasis. *Mol Cell Biol* 21: 175–184.

Facchiano, F., D. D'Arcangelo, A. Riccomi, A. Lentini, S. Beninati, and M. C. Capograssi. 2001. Transglutaminase activity is involved in polyamine-induced programmed cell death. *Exp Cell Res* 271: 118–129.

Fan, H.-F., C.-X. Du, and S.-R. Guo. 2013. Nitric oxide enhances salt tolerance in cucumber seedlings by regulating free polyamine content. *Environ Exp Bot* 86: 52–59.

Farooq, M., S. M. A. Basra, M. Hussain, H. Rehman, and B. A. Saleem. 2007. Incorporation of polyamines in the priming media enhances the germination and early seedling growth in hybrid sunflower (*Helianthus annuus L.*). *Int J Agric Biol* 9: 868–872.

Farooq, M., S. M. A. Basra, H. Rehman, and M. Hussain. 2008. Seed priming with polyamines improves the germination and early seedling growth in fine rice. *J New Seeds* 9: 145–155.

Feldmesser, M., Y. Kress, P. Novikoff, and A. Casadevall. 2000. *Cryptococcus neoformans* is a facultative intracellular pathogen in murine pulmonary infection. *Infect Immun* 68: 4225–4237.

Fellmann, C. and S. W. Lowe. 2014. Stable RNA interference rules for silencing. *Nat Cell Biol* 16: 10–18.

Felschow, D. M., J. MacDiarmids, T. Bardos, R. Wu, P. M. Woster, and C. W. Porter. 1995. Photoaffinity labeling of a cell surface polyamine binding protein. *J Biol Chem* 270: 28705–28711.

Feray, A., A. Hourmant, M. Penot, J. Caroff, and C. Cann-Moisan. 1994. Polyamines and morphogenesis—Effects of methylglyoxal-bis(guanylhydrazone). *Bot Acta* 107: 18–23.

Fere, A., S. Xu, M. K. Montgomery, S. A. Kostas, S. E. Driver, and C. C. Mellow. 1998. Potent and specific genetic interference by double-stranded RNA in *Caenorhabditis elegans*. *Nature* 391: 806–811.

Feuerstein, B. G., N. Pattabiraman, and L. J. Marton. 1990. Molecular mechanics of the interactions of spermine with DNA: DNA bending as a result of ligand binding. *Nucleic Acids Res* 18: 1271–1282.

Filhol, O., C. Cochet, and E. M. Chambaz. 1990. DNA binding activity of casein kinase II. *Biochem Biophys Res Commun* 173: 862–871.

Fiorillo, A., R. Federico, F. Polticelli et al. 2011. The structure of maize polyamine oxidase K300M mutant in complex with the natural substrates provides a snapshot of the catalytic mechanism of polyamine oxidation. *FEBS J* 278: 809–821.

Flamigni, F., A. Facchini, C. Capanni, C. Stefanelli, B. Tantini, and C. M. Caldarera. 1999. p44/42 mitogen-activated protein kinase is involved in the expression of ornithine decarboxylase in leukaemia L1210 cells. *Biochem J* 341: 363–369.

Flink, I. and D. E. Pettijohn. 1975. Polyamines stabilise DNA folds. *Nature* 253: 62–63.

Flores, H. E. and A. W. Galston. 1982. Analysis of polyamines in higher plants by high performance liquid chromatography. *Plant Physiol* 69: 701–706.

Flynn, A. 1996. The role of eIF4 in cell proliferation. *Cancer Surv* 27: 293–310.

Fogel-Petrovic, M., D. L. Kramer, B. Ganis, R. A. Casero, Jr., and C. W. Porter. 1993a. Cloning and sequence analysis of the gene and cDNA encoding mouse spermidine/spermine*N*[1]-acetyltransferase a gene uniquely regulated by polyamines and their analogs. *Biochim Biophys Acta* 1216: 255–264.

Fogel-Petrovic, M., N. W. Shappell, R. J. Bergeron, and C. W. Porter. 1993b. Polyamine and polyamine analog regulation of spermidine/spermine*N*[1]-acetyltransferase in MALME-3M human melonoma cells. *J Biol Chem* 268: 19118–19125.

Folk, J. E., M. H. Park, S. I. Chung, J. Schrode, E. P. Lester, and H. L. Cooper. 1980. Polyamines as physiological substrates for transglutaminases. *J Biol Chem* 255: 3695–3700.

Fontecave, M., M. Atta, and E. Mulliez. 2004. *S*-adenosylmethionine: Nothing goes to waste. *Trends Biochem Sci* 29: 243–249.

Fonzi, W. A. 1989. Regulation of *Saccharomyces cerevisiae* ornithine decarboxylase expression in response to polyamine. *J Biol Chem* 264: 18110–18118.

Fonzi, W. A. and P. S. Sypherd. 1985. Expression of the gene for ornithine decarboxylase of *Saccharomyces cerevisiae* in *Escherichia coli*. *Mol Cell Biol* 5: 161–166.

Fos, M., K. Proano, D. Alavadi, F. Nuez, and J. Carbonell. 2003. Polyamines metabolism is altered in unpollinated parthenocarpic pat-2 tomato ovaries. *Plant Physiol* 131: 359–366.

Foster, S. A. and D. R. Walters. 1990. The effects of polyamine biosynthesis inhibitors on mycelial growth enzyme activity and polyamine levels in the oat-infecting fungus *Pyrenophora avenae*. *J Gen Microbiol* 136: 233–239.

Fraga, M. F., R. Rodriguez, and M. J. Canal. 2002. Genomic DNA methylation–demethylation during aging and reinvigoration of *Pinus radiata*. *Tree Physiol* 22: 813–816.

Francis, S. M., C. A. Taylor, T. Tang, Z. Liu, Q. Zheng, R. Dondero, and J. E. Thompson. 2014. SNS01-T modulation of eIF5A inhibits B-cell cancer progression and synergizes with bortezomib and lenalidomide. *Mol Ther* 22: 1643–1652.

Fraser, A. J. G., R. S. Kamath, P. Zipperten, M. M. Campos, M. Sohrmann, and J. Ahringer. 2000. Functional genomic analysis of *C. elegans* chromosome I by systemic RNA interference. *Nature* 408: 325–330.

Friesen, H., S. R. Hepworth, and J. Segall. 1997. An Ssn6-Tup1-dependent negative regulatory element controls sporulation-specific expression of DIT1 and DIT2 in *Saccharomyces cerevisiae*. *Mol Cell Biol* 17: 123–134.

Friesen, H., J. C. Tanny, and J. Segall. 1998. Spe3, which encodes spermidine synthase, is required for full repression through NRE(DIT) in *Saccharomyces cerevisiae*. *Genetics* 150: 59–73.

Frostesjo, L., I. Holm, B. Grahn, A. W. Page, T. H. Bestor, and O. Heby. 1997. Interference with DNA methyltransferase activity and genome methylation during F9 teratocarcinoma stem cell differentiation induced by polyamine depletion. *J Biol Chem* 272: 4359–4366.

Frugier, M., C. Florentz, M. W. Hosseini, J. M. Lehn, and R. Giege. 1994. Synthetic polyamines stimulate *in vitro* transcription by T7 RNA polymerase. *Nucleic Acids Res* 22: 2784–2790.

Frydman, L., P. C. Rossomando, V. Frydman, C. O. Fernandez, B. Frydman, and K. Samejima. 1992. Interactions between natural polyamines and tRNA: An 15N-NMR analysis. *Proc Natl Acad Sci USA* 89: 9186–9190.

Fuell, C., K. A. Elliott, C. C. Hanfrey, M. Franceschetti, and A. J. Michael. 2010. Polyamine biosynthetic diversity in plants and algae. *Plant Physiol Biochem* 48: 513–520.

Fujimura, K., T. Wright, J. Strnadel et al. 2014. A Hypusine-eIF5A-PEAK1 switch regulates the pathogenesis of pancreatic cancer. *Cancer Res* 74: 6671–6681.

Fujiwara, K., G. Bai, T. Kitagawa, and D. Tsuru. 1998. Immunoelectron microscopic study for polyamines. *J Histochem Cytochem* 46: 1321–1328.

Fukuchi, J., K. Kashiwagi, M. Yamagishi, A. Ishihama, and K. Igarashi. 1995. Decrease in cell viability due to the accumulation of spermidine in spermidine acetyltransferase-deficient mutant of *Escherichia coli*. *J Biol Chem* 270: 18831–18835.

Fuso, A., A. R. Cavallaro, L. Orru, R. F. Buttarelli, and S. Scarpa. 2001. Gene silencing by *S*-adenosylmethionine in muscle differentiation. *FEBS Lett* 508: 337–340.

Gacek, A. and J. Strauss. 2012. The chromatin code of fungal secondary metabolite gene clusters. *Appl Microbiol Biotechnol* 95: 1389–1404.

Gakh, O., P. Cavadini, and G. Isaya. 2002. Mitochondrial processing peptidases. *Biochim Biophys Acta* 1592: 63–77.

Galston, A. W. and R. K. Sawhney. 1990. Polyamines in plant physiology. *Plant Physiol* 94: 406–410.

Gamarnik, A. and R. B. Frydman. 1991. Cadaverine, an essential diamine for the normal root development of germinating soybean (*Glycine max*) seeds. *Plant Physiol* 97: 778–785.

Gamarnik, A., R. B. Frydman, and D. Barreto. 1994. Prevention of infection of soybean seeds by *Colletotrichum truncatum* by polyamine biosynthesis inhibitors. *Phytopathology* 84: 1445–1448.

Gandre, S., Z. Bercovich, and C. Kahana. 2002. Ornithine decarboxylase-antizyme is rapidly degraded through a mechanism that requires functional ubiquitin-dependent proteolytic activity. *Eur J Biochem* 269: 1316–1322.

Ganem, B. 1982. New chemistry of naturally-occurring polyamines. *Acc Chem Res* 15: 290–298.

Garcia, J. R., W. R. Hiatt, J. Peters, and P. S. Sypherd. 1980. *S*-adenosylmethionine levels and protein methylation during morphogenesis of *Mucor racemosus*. *J Bacteriol* 142: 196–201.

García-Jiménez, P., F. García-Maroto, J. A. Garrido-Cárdenas, C. Ferrandiz, and R. R. Robaina. 2009. Differential expression of the ornithine decarboxylase gene during carposporogenesis in the thallus of the red seaweed *Grateloupia imbricata* (Halymeniaceae). *J Plant Physiol* 166: 1745–1754.

Garriz, A., M. C. Dalmasso, F. L. Pieckenstain, and O. A. Ruiz. 2003. The putrescine analogue 1-aminooxy-3-aminopropane perturbs polyamine metabolism in the phytopathogenic fungus *Sclerotinia sclerotiorum*. *Arch Microbiol* 180: 169–175.

Garriz, A., M. E. Gonzalez, M. Marina, O. A. Ruiz, and F. L. Pieckenstain. 2008. Polyamine metabolism during sclerotial development of *Sclerotinia sclerotiorum*. *Mycol Res* 112: 414–422.

Garrod, D. and K. M. Ashworth. 1973. Development of the cellular slime mold *Dictyostelium discoideum*. In *Microbial Differentiation*, eds. J. M. Ashworth and J. E. Smith, 407–435. Cambridge: Cambridge University Press.

Gasch, A. P. 2007. Comparative genomics of the environmental stress response in ascomycete fungi. *Yeast* 24: 961–976.

Gastaut, J. A., G. Tell, P. J. Schechter, D. Maraninchi, B. Mascret, and Y. Carcassonne. 1987. Treatment of acute myeloid leukemia and blastic phase of chronic myeloid leukemia with combined eflornithine (alpha difluoromethylornithine) and methylglyoxal-bis-guanylhydrazone (methyl-GAG). *Cancer Chemother Pharmacol* 20: 344–348.

Ge, C., X. Cui, Y. Wang et al. 2006. BUD2, encoding an *S*-adenosylmethionine decarboxylase, is required for *Arabidopsis* growth and development. *Cell Res* 15: 446–456.

George, C. and M. D. Tsokos. 2011. Systemic lupus erythematosus. *N Engl J Med* 365: 2110–2121.

Gerner, E. W. and F. L. Meyskens, Jr. 2004. Polyamines and cancer: Old molecules, new understanding. *Nat Rev Cancer* 4: 781–792.

Ghochani, B. B. F. N. M. and S. Z. Moosavi-Nejad. 2013. Thermal aggregation of hen egg white lysozyme: Effect of polyamines. *J Paramed Sci* 4: 57–62.

Ghoda, L., T. van Daalen Wetters, M. Macrae, D. Ascherman, and P. Coffino. 1989. Prevention of rapid intracellular degradation of ODC by carboxyl-terminal truncation. *Science* 243: 1493–1495.

Ghoda, L., M. A. Phillips, K. E. Bass, C. C. Wang, and P. Coffino. 1990. Trypanosome ornithine decarboxylase is stable because it lacks sequences found in the carboxyl terminus of the mouse enzyme which target the latter for intracellular degradation. *J Biol Chem* 265: 11823–11826.

Giaever, G., A. M. Chu, L. Ni et al. 2002. Functional profiling of the *Saccharomyces cerevisiae* genome. *Nature* 418: 387–391.

Gicquiaud, L., F. Hennion, and M. A. Esnault. 2002. Physiological comparisons among four related bromus species with varying ecological amplitude: Polyamine and aromatic amine composition in response to salt spray and drought. *Plant Biol* 4: 746–753.

Gill, S. S. and N. Tuteja. 2010. Polyamines and abiotic stress tolerance in plants. *Plant Signal Behav* 5: 26–33.

Ginty, D. D., M. Marlowe, P. H. Pekala, and E. R. Seidel. 1990. Multiple pathways for the regulation of ornithine decarboxylase in intestinal epithelial cells. *Am J Physiol* 258: G454–G460.

Giordano, E., F. Flamigni, C. Guarnieri et al. 2010. Polyamines in cardiac physiology and disease. *Open Heart Failure J* 3: 25–30.

Godderz, D., E. Schafer, R. Palanimurugan, and R. J. Dohmen. 2011. The *N*-terminal unstructured domain of yeast ODC functions as a transplantable and replaceable ubiquitin-independent degron. *J Mol Biol* 407: 354–367.

Goforth, J. B., N. E. Walter, and E. Karatan. 2013. Effects of polyamines on *Vibrio cholerae* virulence properties. *PLoS One* 8:e60765.

Goldemberg, S. H. and I. D. Algranati. 1977. Polyamines and protein synthesis: Studies in various polyamine-requiring mutants of *Escherichia coli*. *Mol Cell Biochem* 16: 71–77.

Goldman, M. E., L. Cregar, D. Nguyen, O. Simo, S. O'Malley, and T. Humphreys. 2006. Cationic polyamines inhibit anthrax lethal factor protease. *BMC Pharmacol* 6: 1–8.

Gonzalez, R. and C. Scazzocchio. 1997. A rapid method for chromatin structure analysis in the filamentous fungus *Aspergillus nidulans*. *Nucleic Acids Res* 25: 3955–3956.

Gonzalez, N. S., C. P. Sanchez, L. Sferco, and I. D. Algranati. 1991. Control of *Leishmania mexicana* proliferation by modulation of polyamine intracellular levels. *Biochem Biophys Res Commun* 180: 797–804.

Gosule, L. C. and J. A. Schellman. 1976. Compact form of DNA induced by spermidine. *Nature* 259: 333–335.

Graff, J. R., A. Benedetti, J. W. Olson, P. Tamez, R. A. Casero, and S. G. Zimmer. 1997. Translation of ODC mRNA and polyamine transport are suppressed in ras-transformed CREFT cells by depleting translation initiation factor 4E. *Biochem Biophys Res Commun* 240: 15–20.

Grant, P. A., L. Duggan, J. Cote et al. 1997. Yeast Gcn5 functions in two multisubunit complexes to acetylate nucleosomal histones: Characterization of an Ada complex and the SAGA (Spt/Ada) complex. *Genes Dev* 11: 1640–1650.

Green, R., C. C. Hanfrey, K. A. Elliott et al. 2011. Independent evolutionary origins of functional polyamine biosynthetic enzyme fusions catalyzing de novo diamine to triamine formation. *Mol Microbiol* 81: 1109–1124.

Gregio, A. P., V. P. Cano, J. S. Avaca, S. R. Valentini, and C. F. Zanelli. 2009. eIF5A has a function in the elongation step of translation in yeast. *Biochem Biophys Res Commun* 380: 785–790.

Grens, A., C. Steglich, R. Pilz, and I. E. Scheffler. 1989. Nucleotide sequence of the Chinese hamster ornithine decarboxylase gene. *Nucleic Acids Res* 17: 10497.

Gritli-Linde, A., I. Holm, and A. Linde. 1995. Localization of *S*-adenosylmethionine decarboxylase in murine tissues by immunohistochemistry. *Eur J Oral Sci* 103: 133–140.

Groppa, M. D., M. L. Tomaro, and M. P. Benavides. 2001. Polyamines as protectors against cadmium or cooper-induced oxidative damage in sunflower leaf discs. *Plant Sci* 161: 481–488.

Groppa, M. D., M. L. Tomaro, and M. P. Benavides. 2007. Polyamines and heavy metal stress: The antioxidant behavior of spermine in cadmium- and copper-treated wheat leaves. *BioMetals* 20: 185–195.

Gruendler, C., Y. Lin, J. Farley, and T. Wang. 2001. Proteasomal degradation of Smad1 induced by bone morphogenetic proteins. *J Biol Chem* 276: 46533–46543.

Grzesiak, M., M. Filek, A. Barbask, B. Kreczmer, and Hartikainen. 2013. Relationships between polyamines, ethylene, osmoprotectans and antioxidant enzymes activities in wheat seedlings after short-term PEG- and NaCl-induced stresses. *Plant Growth Regul* 69: 177–189.

Guevara-Olvera, L., C. Calvo-Mendez, and J. Ruiz-Herrera. 1993. The role of polyamine metabolism in dimorphism of *Yarrowia lipolytica. J Gen Microbiol* 139: 485–493.

Guevara-Olvera, L., B. Xoconostle-Cazares, and J. Ruiz-Herrera. 1997. Cloning and disruption of the ornithine decarboxylase gene of *Ustilago maydis*: Evidence for a role of polyamines in its dimorphic transition. *Microbiology* 43: 2237–2245.

Guevara-Olvera, L., C. Y. Hung, J. J. Yu, and G. T. Cole. 2000. Sequence, expression and functional analysis of the *Coccidioides immitis* ODC (ornithine decarboxylase) gene. *Gene* 242: 437–448.

Guggenberger, C., D. Ilgen, and J. Adamskim. 2007. Functional analysis of cholesterol biosynthesis by RNA interference. *J Steroid Biochem Mol Biol* 104: 105–109.

Gugliucci, A. 2004. Polyamines as clinical laboratory tools. *Clin Chim Acta* 344: 23–35.

Gundogus-Ozcanli, N., C. Sayilir, and W. E. Criss. 1999. Effects of polyamines, polyamine synthesis inhibitors, and polyamine analogs on casein kinase II using Myc oncoprotein as substrate. *Biochem Pharmacol* 58: 251–254.

Guo, X., J. N. Rao, L. Liu, T. T. Zou, D. J. Turner, B. L. Bass, and J. Y. Wang. 2003. Regulation of adherens junctions and epithelial paracellular permeability: A novel function for polyamines. *Am J Physiol* 285: C1174–C1187.

Gupta, R., N. Hamasaki-Katagiri, C. White Tabor, and H. Tabor. 2001. Effect of spermidine on the in vivo degradation of ornithine decarboxylase in *Saccharomyces cerevisiae. Proc Natl Acad Sci U S A* 98: 10620–10623.

Gupta, K., A. Dey, and B. Gupta. 2013. Plant polyamines in abiotic stress responses. *Acta Physiol Plant* 35: 2015–2036.

Guzmán de Peña, D., J. Aguirre, and J. Ruiz-Herrera. 1998. Correlation between the regulation of sterigmatocystin biosynthesis and asexual and sexual sporulation in *Emericella nidulans. A Van Leeuw* 73: 199–205.

Ha, H. C., N. S. Sirisoma, P. Kuppusamy, J. L. Zweier, P. M. Woster, and R. A. Casero, Jr. 1998a. The natural polyamine spermine functions directly as a free radical scavenger. *Proc Natl Acad Sci U S A* 95: 11140–11145.

Ha, H. C., J. D. Yager, P. A. Woster, and R. A. Casero, Jr. 1998b. Structural specificity of polyamines and polyamine analogues in the protection of DNA from strand breaks induced by reactive oxygen species. *Biochem Biophys Res Commun* 244: 298–303.

Hai, Y., J. E. Edwards, M. C. Van Zandt, K. F. Hoffmann, and D. W. Christianson. 2014. Crystal structure of *Schistosoma mansoni* arginase, a potential drug target for the treatment of Schistosomiasis. *Biochemistry* 53: 4671–4684.

Hai, Y., E. J. Kekhoven, M. P. Barrett, and D. W. Christianson. 2015. Crystal structure of an arginase-like protein from *Trypanosoma brucei* that evolved without a binuclear manganese cluster. *Biochemistry* 54: 458–471.

Haider, N., M. L. Eschbach, S. Dias Sde, T. W. Gilberger, R. D. Walter, and K. Luersen. 2005. The spermidine synthase of the malaria parasite *Plasmodium falciparum*: Molecular and biochemical characterisation of the polyamine synthesis enzyme. *Mol Biochem Parasitol* 142: 224–236.

Halmekyto, M., L. Alhonen, J. Wahlfords, R. Sinervirta, T. Eloranta, and J. Janne. 1991. Characterization of a transgenic mouse line over-expressing the human ornithine decarboxylase gene. *Biochemistry* 278: 895–898.

Hamana, K. and S. Matsuzaki. 1985. Distinct difference in the polyamine compositions of bryophyta and pteridophyta. *J Biochem* 97: 1595–1601.

Hamana, K., S. Matsuzaki, M. Niitsu, and K. Samejima. 1994. Distribution of unusual polyamines in aquatic plants and gramineous seeds. *Can J Bot* 72: 1114–1120.

Hamana, K., T. Itoh, Y. Benno, and H. Hayashi. 2008. Polyamine distribution profiles of new members of the phylum bacteroidetes. *J Gen Appl Microbiol* 54: 229–236.

Hamasaki-Katagiri, N., C. W. Tabor, and H. Tabor. 1997. Spermidine biosynthesis in *Saccharomyces cerevisae*: Polyamine requirement of a null mutant of the SPE3 gene (spermidine synthase). *Gene* 187: 35–43.

Hamasaki-Katagiri, N., Y. Katagiri, C. W. Tabor, and H. Tabor. 1998. Spermine is not essential for growth of *Saccharomyces cerevisiae*: Identification of the SPE4 gene (spermine synthase) and characterization of a spe4 deletion mutant. *Gene* 210: 195–201.

Hammond, S., E. Bernstein, D. Beach, and G. J. Hannon. 2000. An RNA-directed nuclease mediates post-transcriptional gene silencing in drosophila cells. *Nature* 404: 293–296.

Hammond, S. M., S. Boettcher, A. A. Caudy, R. Kobayashi, and G. J. Hannon. 2001. Argonaute2, a link between genetic and biochemical analysis of RNAi. *Science* 293: 1146–1150.

Hanfrey, C., S. Sommer, M. J. Mayer, D. Burtin, and A. J. Michael. 2001. *Arabidopsis* polyamine biosynthesis: Absence of ornithine decarboxylase and the mechanism of arginine decarboxylase activity. *Plant J* 27: 551–560.

Hanfrey, C. C., B. M. Pearson, S. Hazeldine et al. 2011. Alternative spermidine biosynthetic route is critical for growth of *Campylobacter jejuni* and is the dominant polyamine pathway in human gut microbiote. *J Biol Chem* 286: 43301–43312.

Hanna, P. 1998. Anthrax pathogenesis and host response. *Curr Topics Microbiol Immunol* 225: 13–35.

Hanzawa, Y., T. Takahashi, and Y. Komeda. 1997. ACL5: An *Arabidopsis* gene required for internodal elongation after flowering. *Plant J* 12: 863–874.

Hanzawa, Y., T. Takahashi, A. J. Michael et al. 2000. ACAULIS5, an *Arabidopsis* gene required for stem elongation, encodes a spermine synthase. *EMBO J* 19: 4248–4256.

Hanzawa, Y., A. Imai, A. J. Michael, Y. Komeda, and T. Takahashi. 2002. Characterization of the spermidine synthase-related gene family in *Arabidopsis thaliana*. *FEBS Lett* 527: 176–180.

Hao, Y. J., H. Kitashiba, C. Honda, K. Nada, and T. Moriguchi. 2005. Expression of arginine decarboxylase and ornithine decarboxylase genes in apple cells and stressed shoots. *J Exp Bot* 56: 1105–1115.

Harinda Champa, W. A., M. I. S. Gill, B. V. C. Mahajan, and S. Bedi. 2014. Exogenous treatment of spermine to maintain quality and extend postharvest life of table grapes (*Vitis vinifera* L.) cv. flame seedless under low temperature storage. *LWT Food Sci Technol* 60: 412–419.

Harold, F. M. 1990. To shape a cell: An inquiry into causes of morphogenesis of microorganisms. *Microbiol Rev* 54: 381–431.

Hart, D., M. Winther, and L. Stevens. 1978. Polyamine distribution and *S*-adenosylmethionine decarboxylase activity in filamentous fungi. *FEMS Microbiol Lett* 3: 173–175.

Hayashi, S., Y. Murakami, and S. Matsufuji. 1996. Ornithine decarboxylase antizyme: A novel type of regulatory protein. *Trends Biochem Sci* 21: 27–30.

Hays, S. M., J. Swanson, and E. U. Selker. 2002. Identification and characterization of the genes encoding the core histones and histone variants of *Neurospora crassa*. *Genetics* 160: 961–973.

Heby, O. 1981. Role of polyamines in the control of cell proliferation and differentiation. *Differentiation* 19: 1–20.

Heby, O. and I. Aqurell. 1971. Observations on the affinity between polyamines and nucleic acids. *Hoppe Seylers Z Physiol Chem* 352: 29–38.

Heby, O. and L. Persson. 1990. Molecular genetics of polyamine synthesis in eukaryotic cells. *Trends Biochem Sci* 15: 153–158.

Heby, O., S. C. Roberts, and B. Ullman. 2003. Polyamine biosynthetic enzymes as drug targets in parasitic protozoa. *Biochem Soc Trans* 3: 415–419.

Heimer, Y. M. and Y. Mizrahi. 1982. Characterization of ornithine decarboxylase of tobacco cells and tomato ovaries. *Biochem J* 201: 373–376.

Hernández, S. M., M. S. Sánchez, and M. N. Schwarcz de Tarlosvsky. 2006. Polyamines as a defense mechanism against lipoperoxidation in *Tripanosoma cruzi*. *Acta Trop* 98: 94–102.

Herrero, A. B., M. C. Lopez, S. Garcia, A. Schmidt, F. Spaltmann, J. Ruiz-Herrera, and A. Dominguez. 1999. Control of filament formation in *Candida albicans* by polyamine levels. *Infect Immun* 67: 4870–4878.

Hibbett, D. S., M. Binder, J. F. Bischoff et al. 2007. A higher-level phylogenetic classification of the fungi. *Mycol Res* 111: 509–547.

Hibino, H., A. Inanobe, K. Furutani, S. Murakami, I. Findlay, and Y. Kurachi. 2010. Inwardly rectifying potassium channels: Their structure, function, and physiological roles. *Physiol Rev* 90: 291–366.

Higashi, K., Y. Terui, A. Suganami, Y. Tamura, K. Nishimura, K. Kashiwagi and K. Igarashi. 2008. Selective structural change by spermidine in the bulged-out region of double-stranded RNA and its effect on RNA function. *J Biol Chem* 283: 32989–32994.

Higgins, D. G., J. D. Thompson, and T. J. Gibson. 1996. Using CLUSTAL for multiple sequence alignments. *Methods Enzymol* 266: 383–402.

Hill, K. and G. E. Schaeller. 2013. Enhancing plant regeneration in tissue culture: A molecular approach through manipulation of cytokinin sensitivity. *Plant Signal Behav* 8: e25709-1–e25709-5.

Hillary, R. A. and A. E. Pegg. 2003. Decarboxylases involved in polyamine biosynthesis and their inactivation by nitric oxide. *Biochem Biophys Acta* 1647: 161–166.

Hobbs, C. A. and S. K. Gilmour. 2006. Role of polyamines in the regulation of chromatin acetylation. In *Polyamine Cell Signaling*, eds. J. Y. Wang and R. A. Casero, 75–89. New Jersey: Humana Press.

Hobbs, C. A., B. A. Paul, and S. K. Gilmour. 2002. Deregulation of polyamine biosynthesis alters intrinsic histone acetyltransferase and deacetylase activities in murine skin and tumors. *Cancer Res* 62: 67–74.

Hobbs, C. A., B. A. Paul, and S. K. Gilmour. 2003. Elevated levels of polyamines alter chromatin in murine skin and tumors without global changes in nucleosome acetylation. *Exp Cell Res* 290: 427–436.

Holtta, E. and P. Pohjanpelto. 1983. Polyamine starvation causes accumulation of cadaverine and its derivatives in a polyamine-dependent strain of Chinese-hamster ovary cells. *Biochem J* 210: 945–948.

Holtta, E., J. Hannonen, J. Pispa, and J. Janne. 1973. Effect of methylglyoxalbis(guanylhydra zone) on polyamine metabolism in normal and regenerating rat liver and rat thymus. *Biochem J* 136: 669–676.

Hougaard, D. M. 1992. Polyamine cytochemistry: Localization and possible functions of polyamines. *Int Rev Cytol* 138: 51–88.

Hougaard, D. M. and L. I. Larsson. 1986. Localization and possible function of polyamines in protein and peptide secreting cells. *Med Biol* 64: 89–94.

Hougaard, D. M., A. M. Del Castillo, and L. I. Larsson. 1988. Endogenous polyamines associate with DNA during its condensation in mammalian tissue. A fluorescence cytochemical and immunocytochemical study of polyamines in fetal rat liver. *Eur J Cell Biol* 45: 311–314.

Hoyt, M. A., M. Broun, and R. H. Davis. 2000. Polyamine regulation of ornithine decarboxylase synthesis in *Neurospora crassa*. *Mol Cell Biol* 20: 2760–2773.

Hoyt, M. A., L. J. Williams-Abbott, J. W. Pitkin, and R. H. Davis. 2000. Cloning and expression of the *S*-adenosylmethionine decarboxylase gene of *Neurospora crassa* and processing of its product. *Mol Gen Genet* 263: 664–673.

Hu, W. W., H. Gong, and E. C. Pua. 2005. The pivotal roles of the plant *S*-adenosylmethionine decarboxylase 5′ untranslated leader sequence in regulation of gene expression at the transcriptional and posttranscriptional levels. *Plant Physiol* 138: 276–286.

Huang, Q., Q. Liu, and Q. Hao. 2005. Crystal structures of Fms1 and its complex with spermine reveal substrate specificity. *J Mol Biol* 348: 951–959.

Hunter, T. 2000. Signaling—2000 and beyond. *Cell* 100: 113–127.

Hussain, S. S., M. Ali, M. Ahmad, and K. H. Siddique. 2011. Polyamines: Natural and engineered abiotic and biotic stress tolerance in plants. *Biotechnol Adv* 29: 300–311.

Hyvonen, M. T., T. A. Keinanen, M. Khomutov et al. 2011. Effects of novel C-methylated spermidine analogs on cell growth via hypusination of eukaryotic translation initiation factor 5A. *Amino Acids* 42: 685–695.

Igarashi, K. and K. Kashiwagi. 1999. Polyamine transport in bacteria and yeast. *Biochem J* 344: 633–642.

Igarashi, K. and K. Kashiwagi. 2000. Polyamines: Mysterious modulators of cellular functions. *Biochem Biophys Res Commun* 271: 559–564.

Igarashi, K. and K. Kashiwagi. 2006. Polyamine modulon in *Escherichia coli*: Genes involved in the stimulation of cell growth by polyamines. *J Biochem* 139: 11–16.

Ikeguchi, Y., M. C. Bewley, and A. E. Pegg. 2006. Aminopropyltransferases: Function, structure and genetics. *J Biochem* 139: 1–9.

Illingworth, C. and A Michael, J. 2012. Plant ornithine decarboxylase is not post-transcriptionally feedback regulated by polyamines but can interact with cytosolic ribosomal protein S15 polypeptide. *Amino Acids* 42: 519–527.

Imai, A., T. Matsuyama, Y. Hanzawa et al. 2004a. Spermidine synthase genes are essential for survival of *Arabidopsis*. *Plant Physiol* 135: 1565–1573.

Imai, A., T. Akiyama, T. Kato, S. Sato, S. Tabata, K. T. Yamamoto, and T. Takahashi. 2004b. Spermine is not essential for survival of *Arabidopsis*. *FEBS Lett* 556: 148–152.

Imanishi, S., K. Hashizume, M. Nakakita et al. 1998. Differential induction by methyl jasmonate of genes encoding ornithine decarboxylase and other enzymes involved in nicotine biosynthesis in tobacco cell cultures. *Plant Mol Biol* 38: 1101–1111.

Incharoensakdi, A., S. Jantaro, W. Raksajit, and P. Maenpaa. 2010. Polyamine in cyanobacteria: Biosynthesis, transport and abiotic stress response. In *Current Research, Technology and Education Topics in Applied Microbiology and Microbial Biotechnology*, ed. A. Méndez-Vila, 23–32. Badajoz: Formatex.

Inderlied, C. B., R. L. Cihlar, and P. S. Sypherd. 1980. Regulation of ornithine decarboxylase during morphogenesis of *Mucor racemosus*. *J Bacteriol* 141: 669–706.

Iqbal, M. and M. Ashraf. 2005. Changes in growth, photosynthethic capacity and ionic relations in spring wheat (*Triticum aestivum* L.) due to pre-sowing seed treatments with polyamines. *Plant Growth Regul* 46: 19–30.

Ito, K., K. Kashiwagi, D. Watanabe, T. Kameji, H. Hayashi, and K. Igarashi. 1990. Influence of the 5′-untranslated region of ornithine decarboxylase mRNA and spermidine on ornithine decarboxylase synthesis. *J Biol Chem* 265: 13036–13041.

Itoh, K., T. Chiba, S. Takahashi et al. 1997. An Nrf/small Maf heterodimer mediates the induction of phase II detoxifying enzymes genes through antioxidant response elements. *Biochem Biophys Res Commun* 236: 313–322.

Ivanov, I. P. and J. F. Atkins. 2007. Ribosomal frameshifting in decoding antizyme mRNAs from yeast and protists to humans: Close to 300 cases reveal remarkable diversity despite underlying conservation. *Nucleic Acids Res* 35: 1842–1858.

Ivanov, I. P., R. F. Gesteland, S. Matsufuji, and J. F. Atkins. 1998. Programmed frameshifting in the synthesis of mammalian antizyme is +1 in mammals, predominantly +1 in fission yeast, but-2 in budding yeast. *RNA* 4: 1230–1238.

Ivanov, I. P., R. F. Gesteland, and J. F. Atkins. 2000. Antizyme expression: A subversion of triplet decoding, which is remarkably conserved by evolution, is a sensor for an autoregulatory circuit. *Nucleic Acids Res* 28: 3185–3196.

Ivanov, I. P., S. Matsufuji, Y. Murakami, R. F. Gesteland, and J. F. Atkins. 2000. Conservation of polyamine regulation by translational frameshifting from yeast to mammals. *EMBO J* 19: 1907–1917.

Ivanov, I. P., R. F. Gesteland, and J. F. Atkins. 2006. Evolutionary specialization of recoding: Frameshifting in the expression of *S. cerevisiae* antizyme mRNA is via an atypical antizyme shift site but is still +1. *RNA* 12: 332–337.

Iwami, K., J. Y. Wang, R. Jain, S. McCormack, and L. R. Johnson. 1990. Intestinal ornithine decarboxylase: Half-life and regulation by putrescine. *Am J Physiol* 258: G308–G315.

Jackson, W. S. 2014. Selective vulnerability to neurodegenerative disease: The curious case of prion protein. *Dis Model Mech* 7: 21–29.

Jackson, L. K., H. B. Brooks, A. L. Osterman, E. J. Goldsmith, and M. A. Phillips. 2000. Altering the reaction specificity of eukaryotic ornithine decarboxylase. *Biochemistry* 39: 11247–11257.

Jakoby, W. B. and J. Fredericks. 1959. Pyrrolidine and putrescine metabolism: Gamma-aminobutyraldehyde dehydrogenase. *J Biol Chem* 234: 2145–2150.

James, T. Y., F. Kauff, C. L. Schoch et al. 2006. Reconstructing the early evolution of fungi using a six-gene phylogeny. *Nature* 443: 818–822.

Jang, E. H., S. A. Park, Y. M. Chi, and K. S. Lee. 2014. Kinetic and structural characterization for cofactor preference of succinic semialdehyde dehydrogenase from *Streptococcus pyogenes*. *Mol Cells* 37: 719–726.

Janicka-Russak, M., K. Kabala, E. Mlodzinska, and G. Klobus. 2010. The role of polyamines in the regulation of the plasma membrane and the tonoplast proton pumps under salt stress. *J Plant Physiol* 167: 261–269.

Janowitz, T., H. Kneifel, and M. Piotrowski. 2003. Identification and characterization of plant agmatine iminohydrolase, the last missing link in polyamine biosynthesis of plants. *FEBS Lett* 544: 258–261.

Jantaro, S., P. Maenpaa, P. Mulo, and A. Incharoensakdi. 2003. Content and biosynthesis of polyamines in salt and osmotically stressed cells of Synechocystis sp. PCC 6803. *FEMS Microbiol Lett* 228: 129–135.

Jauniaux, J.-C., L. A. Urrestarazu, and J–M. Wiame. 1978. Arginine metabolism in *Saccharomyces cerevisiae*: Subcellular localization of the enzymes. *J Bacteriol* 133: 1096–1107.

Jawandha, S. K., M. S. Gill, P. Singh NavPrem, P. S. Gill, and N. Singh. 2012. Effect of post-harvest treatments of putrescine on storage of Mango cv Langra. *Afr J Agric Res* 7: 6432–6436.

Jelsbak, L., L. E. Thomsen, I. Wallrodt, P. R. Jensen, and J. E. Olsen. 2012. Polyamines are require for virulence in *Salmonella enterica* serovar Typhimurium. *PLoS One* 7: e36149.

Jenkinson, C. P., W. W. Grody, and S. D. Cederbaum. 1996. Comparative properties of arginases. *Comp Biochem Physiol Biochem Mol Biol* 114: 107–132.

Jeyapaul, J. and A. K. Jaiswal. 2000. Nrf2 and c-Jun regulation of antioxidant response element (ARE)-mediated expression and induction of glutamylcysteine synthetase heavy subunit gene. *Biochem Pharmacol* 59: 1433–1439.

Jhingran, A., P. K. Padmanabhan, S. Singh et al. 2008. Characterization of the *Entamoeba histolytica* ornithine decarboxylase-like enzyme. *PLoS Negl Trop Dis* 2: e115.

Jian, Q-Q., H-Q. Yang, X-L. Sun, Q. Li, K. Ran, and X-R. Zhang. 2012. Relationship between polyamines metabolism and cell death in roots if *Malus hupehensis* Rehd. under cadmium stress. *J Integr Agric* 11: 1129–1136.

Jimenez-Bremont, J. F., J. Ruiz-Herrera, and A. Dominguez. 2001. Disruption of gene *YlODC* reveals absolute requirement of polyamines for mycelial development in *Yarrowia lipolytica*. *FEMS Yeast Res* 1: 195–204.

Jimenez-Bremont, J. F., E. Hernandez-Lucero, G. Alpuche-Solis, S. Casas-Flores, and A. P. Barba De la Rosa. 2006. Differential distribution of transcripts from genes involved in polyamine biosynthesis in bean plants. *Biol Plantarum* 50: 551–558.

Jin, Y., J. W. Bok, D. Guzman-de-Pena, and N. P. Keller. 2002. Requirement of spermidine for developmental transitions in *Aspergillus nidulans*. *Mol Microbiol* 46: 801–812.

Johnson, T. D. 1996. Modulation of channel function by polyamines. *Trends Pharmacol Sci* 17: 22–27.

Johnson, D. A., R. K. Sharma, K. Allan, R. Ray, and L. R. Johnson. 2004. Immunocytochemical localization of polyamines during attachment and spreading of retinal pigment epithelial and intestinal epithelial cells. *Cell Motil Cytoskeleton* 58: 269–280.

Johnson, K. A., S. Bhushan, A. Stahl, B. M. Hallberg, E. Glaser, and T. Eneqvist. 2006. The closed structure of presequence protease PreP forms a unique 10 000 A°3 chamber proteolysis. *EMBO J* 25: 1977–1986.

Johnson, S. L., D. Jung, M. Forino et al. 2006. Anthrax lethal factor proteases inhibitors: Synthesis, SAR and structure-based 3D QSAR Studies. *J Med Chem* 12: 27–30.

Johnson, L., H. Mulcahy, U. Kanevets, Y. Shi, and S. Lewenza. 2012. Surface-localized spermidine protects the *Pseudomonas aeruginosa* outer membrane from antibiotic treatment and oxidative stress. *J Bacteriol* 194: 813–826.

Jones, M. E. 1985. Conversion of glutamate to ornithine and proline pyrroline-5-carboxylate, a possible modulator of arginine requirements. *J Nutr* 115: 509–515.

Jupe, E. R., J. M. Magill, and C. W. Magill. 1986. Stage-specific DNA methylation in a fungal plant pathogen. *J Bacteriol* 165: 420–423.

Kabir, A. and G. S. Kumar. 2013. Binding of the biogenic polyamines to deoxyribonucleic acids of varying base composition: Base specificity and the associated energetics of the interaction. *PLoS One* 8: 1–13.

Kackar, A. and N. S. Shekhawat. 2007. Plant regeneration through somatic embryogenesis and polyamine levels in cultures of grasses of Thar Desert. *J Cell Mol Biol* 6: 121–127.

Kahana, C. and D. Nathans. 1985. Translational regulation of mammalian ornithine decarboxylase by polyamines. *J Biol Chem* 260: 15390–15393.

Kakkar, R. K., V. K. Rai, and P. K. Nagar. 1997. Polyamine uptake and translocation in plants. *Biol Plantarum* 40: 481–491.

Kakkar, R. K., P. K. Nagar, P. S. Ahuja, and V. K. Rai. 2000. Polyamines and plant morphogenesis. *Biol Plant* 43: 1–11.

Kallio, A., P. McCann, and P. Bey. 1981. DLα(Difluoromethyl)arginine: A potent enzyme activated irreversible inhibitor of bacterial arginine decarboxylase. *Biochemistry* 20: 3163–3166.

Kamatani, N. and D. A. Carson. 1981. Dependence of adenine production upon polyamine synthesis in cultured human lymphoblasts. *Biochim Biophys Acta* 675: 344–350.

Kameji, T. and A. E. Pegg. 1987. Effect of putrescine on the synthesis of *S*-adenosylmethionine decarboxylase. *Biochem J* 243: 285–288.

Kang, H. A. and J. W. B. Hershey. 1994. Effect of initiation factor eIF-5A depletion on protein synthesis and proliferation of *Saccharomyces cerevisiae*. *J Biol Chem* 269: 3934–3940.

Karouzakis, E., R. E. Gay, S. Gay, and M. Neidhart. 2012. Increased recycling of polyamines is associated with global DNA hypomethylation in rheumatoid arthritis synovial fibroblasts. *Arthritis Rheum* 64: 1809–1817.

Kasbek, C., C. H. Yang, and H. A. Fisk. 2010. Antizyme restrains centrosome amplification by regulating the accumulation of Mps1 at centrosomes. *Mol Biol Cell* 21: 3878–3889.

Kashiwagi, K., S. K. Taneja, T. Y. Liu, C. W. Tabor, and H. Tabor. 1990a. Spermidine biosynthesis in *Saccharomyces cerevisiae*. Biosynthesis and processing of a proenzyme form of *S*-adenosylmethionine decarboxylase. *J Biol Chem* 265: 22321–22328.

Kashiwagi, K., Y. Yamaguchi, Y. Sakai, H. Kobayashi, and K. Igarashi. 1990b. Identification of the polyamine-induced protein as a periplasmic oligopeptide binding protein. *J Biol Chem* 265: 8387–8391.

Kashiwagi, K., K. Ito, and K. Igarashi. 1991. Spermidine regulation of ornithine decarboxylase synthesis by a GC-rich sequence of the 5′-untranslated region. *Biochem Biophys Res Commun* 178: 815–822.

Katz, A. and C. Kahana. 1988. Isolation and characterization of the mouse ornithine decarboxylase gene. *J Biol Chem* 263: 7604–7609.

Kay, D. G., R. A. Singer, and G. C. Johnson. 1980. Ornithine decarboxylase activity and cell cycle regulation in *Saccharomyces cerevisiae*. *J Bacteriol* 141: 1041–1046.

Kepczynska, E. 1995. The effects of spermidine biosynthetic inhibitor methyl bis-(guanylhydra zone) on spore germination, growth and ethylene production in *Alternaria consortiale*. *Plant Growth Regul* 16: 263–266.

Ketting, R. F., S. E. Fischer, E. Bernstein, T. Sijen, G. J. Hannon, and R. H. Plasterk. 2001. Dicer functions in RNA interference and in synthesis of small RNA involved in developmental timing in *C. elegans*. *Genes Dev* 15: 2654–2659.

Khan, A. J. and S. C. Minocha. 1989. Biosynthetic arginine decarboxylase in phytopathogenic fungi. *Life Sci* 44: 1215–1222.

Khan, N. A., V. Quemener, and J. P. Moulinoux. 1991. Polyamine membrane transport regulation. *Cell Biol Int Rep* 15: 9–24.

Khan, A. S., Z. Singh, N. A. Abbasi, and E. E. Swinny. 2008. Pre- or post-harvest application of putrescine and low temperature storage affect fruit ripening and quality of "Angelino" plum. *J Sci Food Agric* 88: 1686–1695.

Khan, H. A., K. Ziaf, M. Amjad, and Q. Iqbal. 2012. Exogenous application of polyamines improves germination and early seedling growth of hot pepper. *Chil J Agric Res* 72: 429–433.

Khatri, M. and M. V. Rajam. 2007. Targeting polyamines of *Aspergillus nidulans* by siRNA specific to fungal ornithine decarboxylase gene. *Med Mycol* 45: 211–220.

Khezri, M., A. Talaie, A. Javanshah, and F. Hadavi. 2010. Effect of exogenous application of free polyamines on physiological disorders and yield of "Kaleh-Ghoochi" pistachio shoots (*Pistacia vera* L.). *Sci Hort* 125: 270–276.

Khuhawar, M. Y. and G. A. Qureshi. 2001. Polyamines as cancer markers: Applicable separation methods. *J Chromatogr* 764: 385–407.

Khurana, N., R. K. Saxena, R. Gupta, and M. V. Rajam. 1996. Polyamines as modulators of microcycle conidiation in *Aspergillus flavus*. *Microbiology* 142: 517–523.

Kibe, R., S. Kurihara, Y. Sakai et al. 2014. Upregulation of colonic luminal polyamines produced by intestinal microbiota delays senescence in mice. *Sci Rep* 4: 1–11.

Kim, K. K., L. W. Hung, H. Yakota, R. Kim, and S. H. Kim. 1998. Berkeley structural genomics center. Crystal structures of eukaryotic translation initiation factor 5A from *Methanococcus jannaschii* at 1.8 Å resolution. *Proc Natl Acad Sci U S A* 95: 10419–10424.

Kim, J. K., H. W. Gabel, R. S. Kamath et al. 2005. Functional genomic analysis of RNA interference in *C. elegans*. *Science* 308: 1164–1167.

Kingsbury, J. M., Z. Yang, T. M. Ganous, G. M. Cox, and J. H. McCusker. 2004. Novel chimeric spermidine synthase-saccharopine dehydrogenase gene (*SPE3-LYS9*) in the human pathogen *Cryptococcus neoformans*. *Eukaryot Cell* 3: 752–763.

Kipreos, E. T. and M. Pagano. 2000. The F-box protein family. *Genome Biol* 1: REVIEWS3002.

Klepsch, M. M., M. Kovermann, C. Löw et al. 2011. *Escherichia coli* peptide binding protein OppA has a preference for positively charged peptides. *J Mol Biol* 414: 75–85.

Kobayashi, Y. and K. Horikoshi. 1982. Purification and characterization of extracellular polyamine oxidase produced by *Penicillium* sp. no. PO-1. *Biochim Biophys Acta* 705: 133–138.

Korhonen, V., K. Niiranen, M. Halmekytö et al. 2001. Spermine deficiency resulting from targeted disruption of the spermine synthase gene in embryonic stem cells leads to enhanced sensitivity to antiproliferative drugs. *Mol Pharmacol* 59: 231–238.

Korolev, S., Y. Ikeguchi, T. Skarina, S. Beasley, C. Arrowsmith, and A. Edwards. 2002. The crystal structure of spermidine synthase with a multisubstrate adduct inhibitor. *Nat Struct Biol* 9: 27–31.

Koushesh saba, M., K. Arzani, and M. Barzegar. 2012. Postharvest polyamine application alleviates chilling injury and affects apricot storage ability. *J Agric Food Chem* 60: 8947–8953.

Kozubowski, L. and J. Heitman. 2012. Profiling a killer, the development of *Cryptococcus neoformans*. *FEMS Microbiol Rev* 36: 78–94.

Kramer, G. F., C. Y. Wang, and W. S. Conway. 1991. Inhibition of softening by polyamine application in golden delicious and McIntosh apples. *J Am Soc Hort Sci* 116: 813–817.

Krause, T., K. Luersen, C. Wrenger, T. W. Gilberger, S. Muller, and R. D. Walter. 2000. The ornithine decarboxylase domain of the bifunctional ornithine decarboxylase/*S*-adenosylmethionine decarboxylase of *Plasmodium falciparum*: Recombinant expression and catalytic properties of two different constructs. *Biochem J* 1: 287–292.

Krauss, M., K. Langnaese, K. Richter et al. 2006. Spermidine synthase is prominently expressed in the striatal patch compartment and in putative interneurones of the matrix compartment. *J Neurochem* 97: 174–189.

Krebs, E. G. 1994. The growth of research on protein phosphorylation. *Trends Biochem Sci* 19: 439.

Kristan, K. and T. L. Rizner. 2012. Steroid-transforming enzymes in fungi. *J Steroid Biochem Mol Biol* 129: 79–91.

Kubota, M., E. O. Kajander, and D. A. Carson. 1985. Independent regulation of ornithine decarboxylase and *S*-adenosylmethionine decarboxylase in methylthioadenosine phosphorylase-deficient malignant murine lymphoblasts. *Cancer Res* 45: 3567–3572.

Kuehn, G. D., B. Rodriguez-Garay, S. Bagga, and G. C. Phillips. 1990. Novel occurrence of uncommon polyamines in higher plants. *Plant Physiol* 94: 855–857.

Kumar, A., T. Altabella, M. A. Taylor, and A. F. Tiburcio. 1997. Recent advances in polyamine research. *Trends Plant Sci* 2: 124–130.

Kummasook, A., C. R. Cooper, Jr., A. Sakamoto, Y. Terui, K. Kashiwagi, and N. Vanittanakom. 2013. Spermidine is required for morphogenesis in the human pathogenic fungus, *Penicillium marneffei*. *Fungal Genet Biol* 58–59: 25–32.

Kuramoto, N., K. Inoue, K. Gion, K. Takano, K. Sakata, K. Ogita, and Y. Yoneda. 2003. Endogenous polyamines in murine central and peripheral excitable tissues. *Brain Res* 967: 170–180.

Kurata, H. T., L. J. Marton, and C. G. Nichols. 2006. The polyamine binding site in inward rectifier K+ channels. *J Gen Physiol* 127: 467–480.

Kurian, L., R. Palanimurugan, D. Gödderz, and R. J. Dohmen. 2011. Polyamine sensing by nascent ornithine decarboxylase antizyme stimulates decoding of its mRNA. *Nature* 477: 490–495.

Kusama-Eguchi, K., S. Watanabe, M. Irisawa, K. Watanabe, and K. Igarashi. 1991. Correlation between spermine stimulation of rat liver Ile-tRNA formation and structural change of the acceptor stem by spermine. *Biochem Biophys Res Commun* 177: 745–750.

Kuznetsov, V. V. and N. I. Shevyakova. 2007. Polyamine and stress tolerance of plants. *Plant Stress* 1: 50–71.

Kwak, S.-H. and S. H. Lee. 2002. The transcript-level-independent activation of ornithine decarboxylase in suspension-cultured BY2 cells entering the cell cycle. *Plant Cell Physiol* 43: 1165–1170.

Kyriakidis, D. A. 1983. Effect of plant growth hormones and polyamines on ornithine decarboxylase activity during the germination of barley seeds. *Physiol Plant* 57: 499–504.

Ladenburg, A. and J. Abel. 1888. Über das aethylenimin (Spermin?). *Ber Dtsch Chem Ges* 21: 758–766.

Lagishetty, C. V. and S. R. Naik. 2008. Polyamines: Potential anti-inflammatory agents and their possible mechanism of action. *Indian J Pharmacol* 40: 121–125.

Laitinen, S. I., P. H. Laitinen, and A. Pajunen. 1984. The effect of testosterone on the half-life of ornithine decarboxylase-mRNA in mouse kidney. *Biochem Int* 9: 45–50.

Landau, G., Z. Bercovich, M. H. Park, and C. Kahana. 2010. The role of polyamines in supporting growth of mammalian cells is mediated through their requirement for translation initiation and elongation. *J Biol Chem* 285: 12474–12481.

Landete, J. M., M. E. Arena, I. Pardo, M. C. Manca de Nadra, and S. Ferrer. 2010. The role of two families of bacterial enzymes in putrescine synthesis from agmatine via agmatine deiminase. *Int Microbiol* 13: 169–177.

Landry, J. and R. Sternglanz. 2003. Yeast Fms1 is a FAD-utilizing polyamine oxidase. *Biochem Biophys Res Commun* 303: 771–776.

Lang, G. A., J. D. Early, G. C. Martin, and R. L. Darnell. 1987. Endo-, para-, and eco-dormancy: Physiological terminology and classification for dormancy research. *HortScience* 22: 371–377.

Large, P. J. 1992. Enzymes and pathways of polyamine breakdown in microorganisms. *FEMS Microbiol Rev* 8: 249–262.

Larsson, L. I., L. Morch-Jorgensen, and D. M. Hougaard. 1982. Cellular and subcellular localization of polyamines cytochemical methods providing new clues to polyamine function in normal and neoplastic cells. *Histochemistry* 76: 159–174.

Laube, G. and R. W. Veh. 1997. Astrocytes, not neurons, show most prominent staining for spermidine/spermine-like immunoreactivity in adult rat brain. *Glia* 19: 171–179.

Laube, G., H. G. Bernstein, G. Wolf, and R. W. Veh. 2002. Differential distribution of spermidine/spermine-like immunoreactivity in neurons of the adult rat brain. *J Comp Neurol* 444: 369–386.

Lee, T. M., H. S. Lur, and C. Chu. 1997. Role of abscisic acid in chilling tolerance of rice (*Oriza sativa* L.) seedlings. II. Modulation of free polyamine levels. *Plant Sci* 126: 1–10.

Lee, J., B. Lee, D. Shin, S. S. Kwak, J. D. Bahk, C. O. Lim, and D. J. Yun. 2002. Carnitine uptake by AGP2 in yeast *Saccharomyces cerevisiae* is dependent on Hogl MAP kinase pathway. *Mol Cells* 13: 407–412.

Lee, S. H., Y. J. Yang, K. M. Kim, and B. C. Chung. 2003. Altered urinary profiles of polyamines and endogenous steroids in patients with benign cervical disease and cervical cancer. *Cancer Lett* 201: 121–131.

Lentini, A., B. Provenzano, C. Tabolacci, and S. Beninati. 2009. Protein–polyamine conjugates by transglutaminase 2 as potential markers for antineoplastic screening of natural compounds. *Amino Acids* 36: 701–708.

Lentini, A., C. Tabolacci, B. Provenzano, S. Rossi, and S. Beninati. 2010. Phytochemicals and protein–polyamine conjugates by transglutaminase as chemopreventive and chemotherapeutic tools in cancer. *Plant Physiol Biochem* 48: 627–633.

Leon-Ramirez, C. G., L. Valdes-Santiago, E. Campos-Gongora, L. Ortiz-Castellanos, E. T. Arechiga-Carvajal, and J. Ruiz-Herrera. 2010. A molecular probe for Basidiomycota: The spermidine synthase–saccharopine dehydrogenase chimeric gene. *FEMS Microbiol Lett* 312: 77–83.

Leroy, D., J. K. Heriche, O. Filhol, E. M. Chambaz, and C. Cochet. 1997. Binding of polyamines to an autonomous domain of the regulatory subunit of protein kinase CK2 induces a conformational change in the holoenzyme. A proposed role for the kinase stimulation. *J Biol Chem* 272: 20820–20827.

Levin, E. J., D. A. Kondrashov, G. E. Wesenberg, and G. N. Phillips. 2007. Ensemble refinement of protein crystal structures: Validation and application. *Structure* 15: 1040–1052.

Lewandowski, N. M., S. Ju, M. Verbitsky et al. 2010. Polyamine pathway contributes to the pathogenesis of Parkinson disease. *Proc Natl Acad Sci U S A* 107: 16970–16975.

Li, X. and P. Coffino. 1992. Regulated degradation of ornithine decarboxylase requires interaction with the polyamine-inducible protein antizyme. *Mol Cell Biol* 12: 3556–3562.

Li, G. and D. Reinberg. 2011. Chromatin higher-order structures and gene regulation. *Curr Opin Genet Dev* 21: 175–186.

Libby, P. R. and J. S. Bertram. 1980. Biphasic effect of polyamines on chromatin-bound histone deacetylase. *Arch Biochem Biophys* 201: 359–361.

Lieber, C. S. and L. Packer. 2002. *S*-Adenosylmethionine: Molecular, biological, and clinical aspects—An introduction. *Am J Clin Nutr* 76: 1148S–1150S.

Lindemose, S., P. E. Nielsen, and N. E. Mollegaard. 2005. Polyamines preferentially interact with bent adenine tracts in double-stranded DNA. *Nucleic Acids Res* 33: 1790–1803.

Litchfield, D. W. 2003. Protein kinase CK2: Structure, regulation and role in cellular decisions of life and death. *Biochem J* 369: 1–15.

Liu, J. H. and T. Moriguchi. 2007. Changes in free polyamines and gene expression during preach flower development. *Biol Plantarum* 51: 530–532.

Liu, H., T. R. Cottrell, L. M. Pierini, W. E. Goldman, and T. L. Doering. 2002. RNA interference in the pathogenic fungus *Cryptococcus neoformans*. *Genetics* 160: 463–470.

Liu, B., A. Sutton, and R. Sternglanz. 2005. A yeast polyamine acetyltransferase. *J Biol Chem* 280: 16659–16664.

Liu, H. P., B. J. Yu, W. H. Zhang, and Y. L. Liu. 2005. Effect of osmotic stress on the activity of H+-ATPase and the levels of covalently and non-covalently conjugated polyamines in plasma membrane preparation from wheat seedling roots. *Plant Sci* 168: 1599–1607.

Liu, J. H., H. Inoue, and T. Moriguchi. 2008. Salt stress-mediated changes in free polyamine titers and expression of genes responsible for polyamine biosynthesis of apple in vitro shoots. *Environ Exp Bot* 62: 28–35.

Liu, C., L. Q. Guo, S. Menon et al. 2013a. COP9 signalosome subunit Csn8 is involved in maintaining proper duration of the G1 phase. *J Biol Chem* 288: 20443–20452.

Liu, R., Q. Li, R. Ma, X. Lin, H. Xu, and K. Bi. 2013b. Determination of polyamine metabolome in plasma and urine by ultrahigh performance liquid chromatography–tandem mass spectrometry method: Application to identify potential markers for human hepatic cancer. *Anal Chim Acta* 791: 36–45.

Lockwood, D. H. and L. E. East. 1974. Studies of the insulin-like actions of polyamines on lipid and glucose metabolism in adipose tissue cells. *J Biol Chem* 249: 7717–7722.

Loenen, W. A. 2006. *S*-adenosylmethionine: Jack of all trades and master of everything? *Biochem Soc Trans* 34: 330–333.

Lomozik, L., A. Gasowska, R. Bregier-Jarzebowska, and R. Jastrzab. 2005. Coordination chemistry of polyamines and their interactions in ternary systems including metal ions, nucleosides and nucleotides. *Coord Chem Rev* 249: 2335–2350.

Lopatin, A. N., E. N. Makhina, and C. G. Nichols. 1995. The mechanism of inward rectification of potassium channels: "Long-pore plugging" by cytoplasmic polyamines. *J Gen Physiol* 106: 923–955.

López, M. C., S. García, J. Ruiz-Herrera, and A. Domínguez. 1997. The ornithine decarboxylase gene from *Candida albicans*. Sequence analysis and expression during dimorphism. *Curr Genet* 32: 108–114.

Lopez, J. P., L. M. Fiori, J. A. Gross, B. Labonte, V. Yerko, N. Mechawar, and G. Tureki. 2014. Regulatory role of miRNAs in polyamine gene expression in the prefrontal cortex of depressed suicide completers. *Int J Neuropsychopharmacol* 17: 23–32.

Lu, S. C. 2000. *S*-Adenosylmethionine. *Int J Biochem Cell Biol* 32: 391–395.

Lu, C. D., Y. Itoh, Y. Nakada, and Y. Jiang. 2002. Functional analysis and regulation of the divergent spuABCDEFGH-spuI operons for polyamine uptake and utilization in *Pseudomonas aeruginosa* PAO1. *J Bacteriol* 184: 3765–3773.

Luengwilai, K. and D. M. Beckles. 2013. Effect of low temperature storage on fruit physiology and carbohydrate accumulation in tomato ripening-inhibited mutants. *J Stored Prod Postharvest Res* 4: 35–43.

Luo, J., C.-H. Yu, H. Yu et al. 2013. Cellular polyamines promote amyloid-beta (AB) peptide fibrillation and modulate the aggregation pathways. *ACS Chem Neurosci* 4: 454–462.

Mach, M., M. W. White, M. Neubauer, J. L. Degen, and D. R. Morris. 1986. Isolation of a cDNA clone encoding *S*-adenosylmethionine decarboxylase. Expression of the gene in mitogen-activated lymphocytes. *J Biol Chem* 261: 11697–11703.

Machius, M., C. A. Brautigam, D. R. Tomchick et al. 2007. Structural and biochemical basis for polyamine binding to the Tp0655 lipoprotein of *Treponema pallidum*: Putative role for Tp0655 (Tp0655) as a polyamine receptor. *J Mol Biol* 373: 681–694.

Mackintosh, C. A. and D. R. Walters. 1997. Growth and polyamine metabolism in *Pyrenophora avenae* exposed to cyclohexylamine and norspermidine. *Amino Acids* 13: 347–354.

Mackintosh, C. A. and A. E. Pegg. 2000. Effect of spermine synthase deficiency on polyamine biosynthesis and content in mice and embryonic fibroblasts, and the sensitivity of fibroblasts to 1, 3-bis-(2-chloroethyl)-*N*-nitrosourea. *Biochem J* 351: 439–447.

MacKintosh, E., S. J. Tabrizi, and J. Collinge. 2003. Prion diseases. *J Neurovirol* 9: 183–193.

Madeo, F., T. Eisenberg, S. Buttner, C. Ruckenstuhl, and G. Kroemer. 2010. Spermine: A novel autophagy inducer and longevity elixir. *Autophagy* 6: 160–162.

Magill, J. M. and C. W. Magill. 1989. DNA methylation in fungi. *Dev Genet* 10: 63–69.

Magnes, C., A. Fauland, E. Gander et al. 2014. Polyamines in biological samples: Rapid and robust quantification by solid-phase extraction online-coupled to liquid chromatography–tandem mass spectrometry. *J Chromatogr* 1331: 44–51.

Mala, J., M. Cvikrova, P. Machova, and O. Martincova. 2009. Polyamines during somatic embryo development in Norway spruce (*Piceaabies* [L.]). *J Forest Sci* 55: 75–80.

Malik, A. U., S. C. Tan, and Z. Singh. 2006. Exogenous application of polyamines improves shelf life and fruit quality of mango. *Acta Hort* 699: 291–296.

Malone, T., R. M. Blumenthal, and X. Cheng. 1995. Structure-guided analysis reveals nine sequence motifs conserved among DNA amino-methyltransferases, and suggests a catalytic mechanism for these enzymes. *J Mol Biol* 3: 618–632.

Mangold, U. 2006. Antizyme inhibitor: Mysterious modulator of cell proliferation. *Cell Mol Life Sci* 63: 2095–2101.

Mangold, U. and E. Leberer. 2005. Regulation of all members of the antizyme family by antizyme inhibitor. *Biochem J* 385: 21–28.

Manos, J., J. Arthur, B. Rose et al. 2008. Transcriptome analyses and biofilm-forming characteristics of a clonal *Pseudomonas aeruginosa* from the cystic fibrosis lung. *J Med Microbiol* 57: 1454–1465.

Marchitto, K. S. and A. J. Ferro. 1985. The metabolism of 5′-methylthioadenosine and 5-methylthioribose 1-phosphate in *Saccharomyces cerevisiae*. *J Gen Microbiol* 131: 2153–2164.

Marco, F., R. Alcazar, A. F. Tiburcio, and P. Carrasco. 2011. Interactions between polyamines and abiotic stress pathway responses unraveled by transcriptome analysis of polyamine overproducers. *OMICS* 15: 775–781.

Marcora, M. S., S. Cejas, N. S. González, C. Carrillo, and I. D. Algranati. 2010. Polyamine biosynthesis in Phytomonas: Biochemical characterization of a very unstable ornithine decarboxylase. *Int J Parasitol* 40: 1389–1394.

Maric, S. C., A. Crozat, and O. A. Jänne. 1992. Structure and organization of the human *S*-adenosylmethionine decarboxylase gene. *J Biol Chem* 267: 18915–18923.

Marini, J. C., B. Stoll, I. C. Didelija, and D. G. Burrin. 2012. *De novo* synthesis is the main source of ornithine for citrulline production in neonatal pigs. *Am J Physiol Endocrinol Metab* 303: E1348–E1353.

Marra, M., G. Lombardi, E. Agostinelli et al. 2008. Bovine serum amine oxidase and spm potentiate docetaxel and interferon-alfa effects in inducing apoptosis on human cancer cells through the generation of oxidative stress. *Biochem Biophys Acta* 1783: 2269–2278.

Marshall, M., G. Russo, J. Van Etten, and K. Nickerson. 1979. Polyamines in dimorphic fungi. *Curr Microbiol* 2: 187–190.

Martinez-Pacheco, M. and J. Ruiz-Herrera. 1993. Differential compartmentation of ornithine decarboxylase in cells of *Mucor rouxii*. *J Gen Microbiol* 139: 1387–1394.

Martinez-Pacheco, M., G. Rodriguez, G. Reyna, C. Calvo-Méndez, and J. Ruiz-Herrera. 1989. Inhibition of the yeast-mycelial transition and the phorogenesis of *Mucorales* by diamino butanone. *Arch Microbiol* 151: 10–14.

Martin-Tanguy, J. 2001. Metabolism and function of polyamines in plants: Recent development (new approaches). *Plant Growth Regul* 34: 135–148.

Martin-Tanguy, J. 2006. Conjugated polyamines and reproductive development: Biochemical, molecular and physiological. *Physiol Plantarum* 100: 675–688.

Mato, J. M., L. Alvarez, P. Ortiz, and M. A. Pajares. 1997. *S*-adenosylmethionine synthesis: Molecular mechanisms and clinical implications. *Pharmacol Ther* 73: 265–280.

Matsufuji, S., T. Matsufuji, Y. Miyazaki, Y. Murakami, J. F. Atkins, R. F. Gesteland, and S. Hayashi. 1995. Autoregulatory frameshifting in decoding mammalian ornithine decarboxylase antizyme. *Cell* 80: 51–60.

Matsumoto, M., S. Kurihara, R. Kibe, H. Aschidam, and Y. Benno. 2011. Longevity in mice is promoted by probiotic-induced suppression of colonic senescence dependent on upregulation of gut bacterial polyamine production. *PLoS One* 6: e23652.

Matthews, H. R. 1993. Polyamines, chromatin structure and transcription. *Bioessays* 15: 561–566.

Mattoo, A. K. and A. K. Handa. 2008. Higher polyamines restore and enhance metabolic memory in ripening fruit. *Plant Sci* 174: 386–393.

Mayer, M. J. and A. J. Michael. 2003. Polyamine homeostasis in transgenic plants overexpressing ornithine decarboxylase induces ornithine limitation. *J Biochem* 134: 765–772.

McCann, P. P. and A. E. Pegg. 1992. Ornithine decarboxylase as an enzyme target for therapy. *Pharmacol Ther* 54: 195–215.

McCormack, S. A., J. Y. Wang, M. J. Viar, L. Taque, P. J. Davies, and L. R. Jhonson. 1994. Polyamines influence transglutaminase activity and cell migration in two cell lines. *Am J Physiol* 267: C706–C714.

McCullough, J. L., G. D. Weinstein, M. G. Rosenblum, and J. J. Jenkins. 1983. Percutaneous penetration of methylglyoxalbis(guanylhydrazone): Effects on hairless mouse epidermis *in vivo*. *J Invest Dermatol* 81: 388–392.

Meers, P., K. Hong, J. Bentz, and D. Papahadjopoulos. 1986. Spermine as a modulator of membrane fusion: Interactions with acidic phospholipids. *Biochemistry* 25: 3109–3118.

Meijer, H. A. and A. A. Thomas. 2002. Control of eukaryotic protein synthesis by upstream open reading frames in the 5′-untranslated region of an mRNA. *Biochem J* 367: 1–11.

Meister, G. 2013. Argonaute proteins: Functional insights and emerging roles. *Nat Rev Genet* 14: 447–459.

Mémin, E., M. Hoque, R. J. Mohit et al. 2013. Blocking eIF5A modification in cervical cancer cells alters the expression of cancer-related genes and suppresses cell proliferation. *Cancer Res* 74: 552–562.

Mendgen, K., M. Hahn, and H. Deising. 1996. Morphogenesis and mechanisms of penetration by plant pathogenic fungi. *Annu Rev Phytopathol* 34: 367–386.

Merali, S., M. Saric, K. Chin, and A. B. Clarkson, Jr. 2000. Effect of a bis-benzyl polyamine analogue on *Pneumocystis carinii*. *Antimicrob Agents Chemother* 44: 337–343.

Meskaoui, A. E. and F. M. Trembaly. 2009. Effects of exogenous polyamines and inhibitors of polyamine biosynthesis on endogenous free polyamine contents and the maturation of white spruce somatic embryos. *Afr J Biotechnol* 8: 6807–6816.

Metcalf, B. W., P. Bey, C. Danzin, M. J. Jung, P. Casara, and J. P. Vevert. 1978. Catalytic irreversible inhibition of mammalian ornithine decarboxylase by substrate and product analogues. *J Am Chem Soc* 100: 2251–2253.

Meyskens, F. L. and E. W. Gerner. 1999. Development of difluoromethylornithine (DFMO) as a chemoprevention agent. *Clin Cancer Res* 5: 945–951.

Michael, A., J. M. Furze, J. C. Rhodes, and D. Burtin. 1996. Molecular cloning and functional identification of a plant ornithine decarboxylase cDNA. *Biochem J* 314: 241–248.

Minguet, E. G., F. Vera-Sirera, A. Marina, J. Carbonell, and M. A. Blazquez. 2008. Evolutionary diversification in polyamine biosynthesis. *Mol Biol Evol* 25: 2119–2128.

Minocha, R., R. Majumdar, and S. C. Minocha. 2014. Polyamines and abiotic stress in plants: A complex relationship. *Fron Plant Sci* 5: 1–17.

Minois, N., D. Carmona-Gutierrez, and F. Madeo. 2011. Polyamines in aging and disease. *Aging (Albany NY)* 3: 716–732.

Mirdehghan, S. H., M. Rahemi, S. Castillo, D. Martínez-Romero, M. Serrano, and D. Valero. 2007. Pre-storage application of polyamines by pressure or immersion improves shelf-life of pomegranate stored at chilling temperature by increasing encogenous polyamine levels. *Postharvest Biol Technol* 44: 26–33.

Mitchell, T. G. and J. R. Perfect. 1995. Cryptococcosis in the era of AIDS—100 years after the discovery of *Cryptococcus neoformans*. *Clin Microbiol Rev* 8: 515–548.

Mitchell, J. L., G. G. Judd, A. Bareyal-Leyser, and S. Y. Ling. 1994. Feedback repression of polyamine transport is mediated by antizyme in mammalian tissue-culture cells. *Biochem J* 299: 19–22.

Mitchell, D. H., T. C. Sorrell, A. M. Allworth et al. 1995. *Cryptococcal* disease of the CNS in immunocompetent hosts: Influence of cryptococcal variety on clinical manifestations and outcome. *Clin Infect Dis* 20: 611–616.

Miyake, M., T. Minami, M. Hirota et al. 2006. Novel oral formulation safety improving intestinal absorption of poorly absorbable drugs: Utilization of polyamines and bile acids. *J Control Release* 111: 27–34.

Mize, G. J. and D. R. Morris. 2001. A mammalian sequence-dependent upstream open reading frame mediates polyamine-regulated translation in yeast. *RNA* 7: 374–381.

Mize, G. J., H. Ruan, J. J. Low, and D. R. Morris. 1998. The inhibitory upstream open reading frame from mammalian *S*-adenosylmethionine decarboxylase mRNA has a strict sequence in critical positions. *J Biol Chem* 271: 32500–32505.

Moazeni, M., Nabili, M., H. Badali, and M. Abastabar. 2014. RNAi technology: A novel approach against fungal infections. *Res Mol Med* 2: 1–10.

Moinard, C., L. Cynober, and J. P. De Bandt. 2005. Polyamines: Metabolism and implications in human diseases. *Clin Nutr* 24: 184–197.

Moore, D. 1998. *Fungal Morphogenesis Development and Cell Biology Series*. Cambridge: Cambridge University Press.

Moore, K. S., S. Wehrli, H. Roder, M. Rogers, J. N. Forrest, Jr., D. McCrimmon, and M. Zasloff. 1993. Squalamine: An aminosterol antibiotic from the shark. *Proc Natl Acad Sci U S A* 90: 1354–1358.

Mora, J., R. Tarrab, J. Martuscelli, and G. Soberón. 1965. Characteristics of arginases from ureotelic and non-ureotelic animals. *Biochem J* 96: 588–594.

Morgan, D. M. 1998. Polyamine oxidases enzymes of unknown function? *Biochem Soc Trans* 26: 586–591.

Morgan, J. E., J. W. Blankenship, and H. R. Matthews. 1987. Polyamines and acetylpolyamines increase the stability and alter the conformation of nucleosome core particles. *Biochemistry* 26: 3643–3649.

Morimoto, N., W. Fukuda, N. Nakajima et al. 2010. Dual biosynthesis pathway for longer-chain polyamines in the hyperthermophilic archaeon *Thermococcus kodakarensis*. *J Bacteriol* 192: 4991–5001.

Morris, S. M., Jr. 2002. Regulation of enzymes of the urea cycle and arginine metabolism. *Annu Rev Nutr* 22: 87–105.

Morris, D. R. and A. P. Geballe. 2000. Upstream open reading frames as regulators of mRNA translation. *Mol Cell Biol* 20: 8635–8642.

Moschou, P. N. and K. A. Roubelakis-Angelakis. 2011. Characterization, assay, and substrate specificity of plant polyamine oxidases. *Methods Mol Biol* 720: 183–194.

Moshier, J. A., J. D. Gilbert, M. Skinca, J. Dosescu, K. M. Almodovar, and G. D. Luk. 1990. Isolation and expression of a human ornithine decarboxylase gene. *J Biol Chem* 265: 4884–4892.

Mueller, E., A. Bailey, A. Corran, A. J. Michael, and P. Bowyer. 2001. Ornithine decarboxylase knockout in *Tapesia yallundae* abolishes infection plaque formation *in vitro* but does not reduce virulence toward wheat. *Mol Plant Microbe Interact* 14: 1303–1311.

Muller, M., M. Cleef, G. Rohn, P. Bonnekoh, A. E. Pajunen, H. G. Bernstein, and W. Paschen. 1991. Ornithine decarboxylase in reversible cerebral ischemia: An immunohistochemical study. *Acta Neuropathol* 83: 39–45.

Munder, M. 2009. Arginase: An emerging key player in the mammalian immune system. *Br J Pharmacol* 158: 638–651.

Murai, N., Y. Murakami, and S. Matsufuji. 2003. Identification of nuclear export signals in antizyme-1. *J Biol Chem* 278: 44791–44798.

Murai, N., A. Shimizu, Y. Murakami, and S. Matsufuji. 2009. Subcellular localization and phosphorylation of antizyme 2. *J Cell Biochem* 108: 1012–1021.

Murakami, Y., S. Matsufuji, M. Nishiyama, and S. Hayashi. 1989. Properties and fluctuations in vivo of rat liver antizyme inhibitor. *Biochem J* 259: 839–845.

Murakami, Y., S. Matsufuji, T. Kameji et al. 1992. Ornithine decarboxylase is degraded by the 26S proteasome without ubiquitination. *Nature* 360: 597–599.

Murray-Stewart, T., Y. Wang, A. Goodwin, A. Hacker, A. Meeker, and R. A. Casero, Jr. 2008. Nuclear localization of human spermine oxidase isoforms—Possible implications in drug response and disease etiology. *FEBS J* 275: 2795–2806.

Muth, A., M. Madan, J. J. Archer, N. Ocampo, L. Rodriguez, and O. Phanstiel, 4th. 2014. Polyamine transport inhibitors: Design, synthesis, and combination therapies with difluoromethylornithine. *J Med Chem* 23: 348–363.

Mutlu, F. and S. Bozcuk. 2007. Salinity-induced changes of free and bound polyamine levels in sunflower (*Helianthus Annuus* L.) roots differing in salt tolerance. *Pak J Bot* 39: 1097–1102.

Myers, D. P., L. K. Jackson, V. G. Ipe, G. E. Murphy, and M. A. Phillips. 2001. Long-range interactions in the dimer interface of ornithine decarboxylase are important for enzyme function. *Biochemistry* 40: 13230–13236.

N'Soukpoe-Kossi, C. N., A. A. Ouameur, T. Thomas, A. Shirahata, T. J. Thomas, and H. A. Tajmir-Riahi. 2008. DNA interaction with antitumor polyamine analogues: A comparison with biogenic polyamines. *Biomacromolecules* 9: 2712–2718.

Nahlik, K., M. Dumkow, O. Bayram et al. 2010. The COP9 signalosome mediates transcriptional and metabolic response to hormones, oxidative stress protection and cell wall rearrangement during fungal development. *Mol Microbiol* 78: 964–979.

Nakada, Y. and Y. Itoh. 2003. Identification of the putrescine biosynthetic genes in *Pseudomonas aeruginosa* and characterization of agmatine deiminase and *N*-carbamoylputrescine amidohydrolase of the arginine decarboxylase pathway. *Microbiology* 149: 707–714.

Namy, O., A. Galopier, C. Martini, S. Matsufuji, C. Fabret, and J.-P. Rousset. 2008. Epigenetic control of polyamines by the prion [PSI+]. *Nat Cell Biol* 10: 1069–1075.

Newman, R. M., A. Mobascher, U. Mangold, C. Koike, S. Schmidt, D. Finley, and B. R. Zetter. 2004. Antizyme targets cyclin D1 for degradation. A novel mechanism for cell growth repression. *J Biol Chem* 279: 41504–41511.

Newton, G. L., J. A. Aguilera, J. F. Ward, and R. C. Fahey. 1996. Polyamine-induced compaction and aggregation of DNA—A major factor in radioprotection of chromatin under physiological conditions. *Radiat Res* 145: 776–780.

Nickerson, W. J. and S. Bartnicki-García. 1964. Biochemical aspects of morphogenesis in algae and fungi. *Annu Rev Plant Physiol* 15: 327–344.

Nickerson, K. W. and L. C. Lane. 1977. Polyamine content of several RNA plant viruses. *Virology* 81: 455–459.

Nickerson, K. W., L. D. Dunkle, and J. L. Van Etten. 1977. Absence of spermine in filamentous fungi. *J Bacteriol* 129: 173–176.

Nieves, N., M. E. Martinez, R. Castillo, M. A. Blanco, and J. L. Gonzalez-Olmedo. 2001. Effect of abscisic acid and jasmonic acid on partial dessication of encapsulated somatic embryos of sugarcane. *Plant Cell Tissue Organ Cult* 65: 15–21.

Nikolaou, E., I. Agrafioti, M. Stumpf, J. Quinn, I. Stansfield, and A. J. Brown. 2009. Phylogenetic diversity of stress signalling pathways in fungi. *BMC Evol Biol* 9: 44.

Nilsson, J., A. Gritli-Linde, and O. Heby. 2000. Skin fibroblasts from spermine synthase-deficient hemizygous gyro male (Gy/Y) mice overproduce spermidine and exhibit increased resistance to oxidative stress but decreased resistance to UV radiation. *Biochem J* 352: 381–387.

Nishibori, N. and T. Nishijima. 2003. Changes in polyamine levels during growth of a red-tide causing phytoplankton *Chattonellaantiqua. Eur J Phycol* 39: 51–55.

Nishikawa, M., T. Hagishita, H. Yurimoto, N. Kato, Y. Sakai, and T. Hatanaka. 2000. Primary structure and expression of peroxisomal acetylspermidine oxidase in the methylotrophic yeast *Candida boidinii. FEBS Lett* 476: 150–154.

Nishimura, K., S. B. Lee, J. H. Park, and M. H. Park. 2011. Essential role of eIF5A-1 and deoxyhypusine synthase in mouse embryonic development. *Amino Acids* 42: 703–710.

Njuguna, J. T., M. Nassar, A. Hoerauf, and A. E. Kaiser. 2006. Cloning, expression and functional activity of deoxyhypusine synthase from *Plasmodium vivax. BMC Microbiol* 6: 1–7.

Nordberg, H., M. Cantor, S. Dusheyko et al. 2014. The genome portal of the department of energy joint genome institute: 2014 updates. *Nucleic Acids Res* 42: D26–D31.

Ober, D. and T. Hartmann. 1999. Deoxyhypusine synthase from tobacco. cDNA isolation, characterization, and bacterial expression of an enzyme with extended substrate specificity. *J Biol Chem* 274: 32040–32047.

Obregon, A., S. Monzalvo, C. Calvo-Mendez, and J. Ruiz-Herrera. 1990. Ultraestructural and chemical alteration in germinating spores of *Mucor rouxii* (Zygomycetes), induced by two compounds which inhibit their developmental pattern. *Crypt Bot* 1: 323–331.

Ogita, S. H., H. Uefuji, Y. Yamaguchi, N. Koizumi, and H. Sano. 2003. Producing decaffeinated coffee plants. *Nature* 423: 823.

Ohe, M., M. Kobayashi, M. Niitsu, N. Bagni, and S. Matsuzaki. 2005. Analysis of polyamine metabolism in soybean seedlings using 15N-labelled putrescine. *Phytochemistry* 66: 523–528.

Oredsson, S. M., D. S. Friend, and L. J. Marton. 1983. Changes in mitochondrial structure and function in 9L rat brain tumor cells treated in vitro with alfa-difluoromethylornithine, a polyamine biosynthesis inhibitor. *Proc Natl Acad Sci U S A* 80: 780–784.

Oshima, T. 2007. Unique polyamines produced by an extreme thermophile, *Thermus thermophilus. Amino Acids* 33: 367–372.

Oshima, T., T. Moriya, and Y. Terui. 2011. Identification, chemical synthesis, and biological functions of unusual polyamines produced by extreme thermophiles. In *Polyamines. Methods and Protocols*, eds. A. E. Pegg and R. A. J. Casero, 81–111. New York: Humana Press.

Osterman, A., N. V. Grishin, L. N. Kinch, and M. A. Phillips. 1994. Formation of functional cross-species heterodimers of ornithine decarboxylase. *Biochemistry* 33: 13662–13667.

Otzen, C., S. Muller, I. D. Jacobsen, and M. Brock. 2013. Phylogenetic and phenotypic characterisation of the 3-ketoacyl-CoA thiolase gene family from the opportunistic human pathogenic fungus *Candida albicans*. *FEMS Yeast Res* 13: 553–564.

Ouameur, A. A. and H. A. Tajmir-Riahi. 2004. Structural analysis of DNA interactions with biogenic polyamines and cobalt (III) hexamine studied by Fourier transform infrared and capillary electrophoresis. *J Biol Chem* 279: 42041–42054.

Ouameur, A. A., P. Bourassa, and H. A. Tajmir-Riahi. 2010. Probing tRNA interaction with biogenic polyamines. *RNA* 16: 1968–1979.

Pabo, C. O. and R. T. Sauer. 1992. Transcription factors: Structural families and principles of DNA recognition. *Annu Rev Biochem* 61: 1053–1095.

Palanimurugan, R., H. Scheel, K. Hofmann, and R. J. Dohmen. 2004. Polyamines regulate their synthesis by inducing expression and blocking degradation of ODC antizyme. *EMBO J* 23: 4857–4867.

Palma, F., F. Carvajal, M. Jamilena, and D. Garrido. 2014. Contribution of polyamines and other related metabolites to the maintenance of zucchini fruit quality during cold storage. *Plant Physiol Biochem* 82: 161–171.

Palma, F., F. Carvajal, J. M. Ramos, M. Jamilena, and D. Garrido. 2015. Effect of putrescine application on maintenance of zucchini fruit quality during cold storage: Contribution of GABA shunt and other related nitrogen metabolites. *Postharvest Biol Technol* 99: 131–140.

Panagiotidis, C. A., S. Artandi, K. Calame, and S. J. Silverstein. 1995. Polyamines alter sequence-specific DNA–protein interactions. *Nucleic Acids Res* 23: 1800–1809.

Pandey, S., S. A. Ranade, P. K. Nagar, and N. Kumar. 2000. Role of polyamines and ethylene as modulators of plant senescence. *J Biosci* 25: 291–299.

Pannifer, A. D., T. Y. Wong, R. Schwarzenbacher et al. 2001. Crystal structure of the anthrax lethal factor. *Nature* 414: 229–233.

Paquette, M. A., A. A. Martinez, T. Macheda, C. K. Meshul, S. W. Johnson, S. P. Berger, and A. Giuffrida. 2012. Anti-diskinetic mechanisms of amantadine and dextromethorphan in the 6-OHDA rat model of Parkinson's disease: Role of NMDA vs. 5-HT1A receptors. *Eur J Neurosci* 36: 3224–3234.

Park, M. H. 2006. The post-translational synthesis of a polyamine-derived amino acid, hypusine, in the eukaryotic translation initiation factor 5A (eIF5A). *J Biochem* 139: 161–169.

Park, M. H. and E. C. Wolff. 1988. Cell-free synthesis of deoxyhypusine. Separation of protein substrate and enzyme and identification of 1,3-diaminopropane as a product of spermidine cleavage. *J Biol Chem* 263: 15264–15269.

Park, M. H., H. L. Cooper, and J. E. Folk. 1981. Identification of hypusine, an unusual amino acid, in a protein from human lymphocytes and of spermidine as its biosynthetic precursor. *Proc Natl Acad Sci U S A* 78: 2869–2873.

Park, M. H., E. C. Wolff, and J. E. Folk. 1993a. Hypusine: Its post-translational formation in eukaryotic initiation factor 5A and its potential role in cellular regulation. *Biofactors* 4: 95–104.

Park, M. H., E. C. Wolff, and J. E. Folk. 1993b. Is hypusine essential for eukaryotic cell proliferation? *Trends Biochem Sci* 18: 475–479.

Park, M. H., Y. B. Lee, and Y. A. Joe. 1997. Hypusine is essential for eukaryotic cell proliferation. *Biol Signals* 6: 115–123.

Park, J. H., L. Aravind, E. C. Wolff, J. Kaevel, Y. S. Kim, and M. H. Park. 2006. Molecular cloning, expression, and structural prediction of deoxyhypusine hydroxylase: A HEAT-repeat-containing metalloenzyme. *Proc Natl Acad Sci U S A* 103: 51–56.

Park, B. J., K. A. Wannemuehler, B. J. Marston, N. Govender, P. G. Pappas, and T. M. Chiller. 2009. Estimation of the current global burden of cryptococcal meningitis among persons living with HIV/AIDS. *AIDS* 23: 525–530.

Parvin, S., O. R. Lee, G. Sathiyaraj, A. Khorolragchaa, Y.-J. Kim, and D.-C. Yang. 2014. Spermidine alleviates the growth of saline-stressed ginseng seedlings through antioxidative defense system. *Gene* 537: 70–78.

Paschalidis, K. A. and K. A. Roubelakis-Angelakis. 2005. Spatial and temporal distribution of polyamine levels and polyamine anabolism in different organs/tissues of the tobacco plant. Correlations with age, cell division/expansion, and differentiation. *Plant Physiol* 138: 142–152.

Pastre, D., O. Pietrement, F. Landousy et al. 2006. A new approach to DNA bending by polyamines and its implication in DNA condensation. *Eur Biophys J* 35: 214–223.

Patel, A. R. and J. Y. Wang. 1997. Polyamines modulate transcription but not posttranscription of c-myc and c-jun in IEC-6 cells. *Am J Physiol* 273: C1020–C1029.

Patel, M. M. and T. J. Anchordoquy. 2006. Ability of spermine to differentiate between DNA sequences—Preferential stabilization of A-tracts. *Biophys Chem* 122: 5–15.

Patel, P. H., M. Costa-Mattioli, K. L. Schulze, and H. J. Bellen. 2009. The drosophila deoxyhypusine hydroxylase homologue nero and its target eIF5A are required for cell growth and the regulation of autophagy. *J Cell Biol* 185: 1181–1194.

Paulus, T. J., P. Kiyono, and R. H. Davis. 1982. Polyamine-deficient *Neurospora crassa* mutants and synthesis of cadaverine. *J Bacteriol* 152: 291–297.

Paulus, T. J., C. L. Cramer, and R. H. Davis. 1983. Compartmentation of spermidine in *Neurospora crassa*. *J Biol Chem* 258: 8608–8612.

Pedroso, M. C., N. Primikirios, A. Roubelakis-Angelakis, and M. S. Pasi. 1997. Free and conjugated polyamines in embryogenic and non embryogenic leaf regions of Camellia leaves before and during direct somatic embryogenesis. *Physiol Plant* 101: 213–219.

Pegg, A. E. 1984. *S*-adenosylmethionine decarboxylase: A brief review. *Cell Biochem Funct* 2: 11–15.

Pegg, A. E. 1986. Recent advances in the biochemistry of polyamines in eukaryotes. *Biochem J* 234: 249–262.

Pegg, A. E. 1987. The use of inhibitors to study the biochemistry and molecular biology of polyamine biosynthesis and uptake. In *Inhibition of Polyamine Metabolism. Biological Significance and Basis for New Therapies*, eds. P. P. McCann, A. E. Pegg, and A. Sjoerdsma, 107–119. Orlando, FL: Academic Press.

Pegg, A. E. 1988. Polyamine metabolism and its importance in neoplastic growth and a target for chemotherapy. *Cancer Res* 48: 759–774.

Pegg, A. E. 2008. Spermidine/spermine-*N*(1)-acetyltransferase: A key metabolic regulator. *Am J Physiol Endocrinol Metab* 294: E995–E1010.

Pegg, A. E. 2009. *S*-Adenosylmethionine decarboxylase. *Essays Biochem* 4: 25–45.

Pegg, A. E. and H. G. Williams-Ashman. 1969. On the role of *S*-adenosyl-L-methionine in the biosynthesis of spermidine by rat prostate. *J Biol Chem* 244: 682–693.

Pegg, A. E. and S. McGill. 1979. Decarboxylation of ornothine and lysine in rat tissues. *Biochem Biophys Acta* 568: 416–427.

Pegg, A. E. and P. P. McCann. 1982. Polyamine metabolism and function. *Am J Physiol* 243: C212–C221.

Pegg, A. E. and H. G. Williams-Ashman. 1987. Pharmacologic interference with enzymes of polyamine biosynthesis and of 5′-methylthioadenosine metabolism. In *Inhibition of Polyamine Metabolism: Biological Significance and Basis for New Therapies*, eds. P. P. McCann, A. E. Pegg, and A. Sjoerdsma, 33–48. San Diego, CA: Academic Press.

Pegg, A. E. and P. P. McCann. 1992. *S*-adenosylmethionine decarboxylase as an enzyme target for therapy. *Pharmacol Ther* 56: 359–377.

Pegg, A. E. and A. J. Michael. 2010. Spermine synthase. *Cell Mol Life Sci* 67: 113–121.

Pegg, A. E., R. Madhubala, T. Kameji, and R. J. Bergeron. 1988. Control of ornithine decarboxylase activity in a-difluoromethylornithine-resistant L1210 cells by polyamines and synthetic analogues. *J Biol Chem* 263: 11008–11014.

Pegg, A. E., R. Pakala, and R. J. Bergeron. 1990. Induction of spermidine/spermine N^1,N^{11}-bis(ethyl)norspermine and related compounds. *Biochem J* 267: 331–338.

Pegg, A. E., H. Xiong, D. J. Feith, and L. M. Shantz. 1998. *S*-adenosylmethionine decarboxylase: Structure, function and regulation by polyamines. *Biochem Soc Trans* 26: 580–586.

Pérez-Vicente, A., D. Martínez-Romero, A. Carbonell, M. Serrano, F. Riquelme, F. Guillén, and D. Valero. 2002. Role of polyamines in extending shelf life and the reduction of mechanical damage during plum (*Prunus salicina* Lindl.). *Postharvest Biol Technol* 25: 25–32.

Perfect, J. R., B. Wong, Y. C. Chang, K. J. Kwon-Chung, and P. R. Williamson. 1998. *Cryptococcus neoformans*: Virulence and host defences. *Med Mycol* 36: 79–86.

Persson, L., I. Holm, and O. Heby. 1986. Translational regulation of ornithine decarboxylase by polyamines. *FEBS Lett* 205: 175–178.

Persson, L., A. Jeppsson, and S. Nasizadeh. 2003. Turnover of trypanosomal ornithine decarboxylases. *Biochem Soc Trans* 31: 411–414.

Pfaller, M. A., J. Riley, and T. Gerarden. 1990. Polyamine depletion and growth inhibition of *Cryptococcus neoformans* by alpha-difluoromethylornithine and cyclohexylamine. *Mycopathology* 112: 27–32.

Phillips, M. A., P. Coffino, and C. C. Wang. 1987. Cloning and sequencing of the ornithine decarboxylase gene from *Trypanosoma brucei* implications for enzyme turnover and selective difluoromethylornithine inhibition. *J Biol Chem* 262: 8721–8727.

Pieckenstain, F. L., A. Garriz, E. M. Chornomaz, D. H. Sanchez, and O. A. Ruiz. 2001. The effect of polyamine biosynthesis inhibition on growth and differentiation of the phytopathogenic fungus *Sclerotinia sclerotiorum. Anton Leeuw* 80: 245–253.

Pistocchi, R., F. Keller, N. Bagni, and P. Matile. 1988. Transport and subcellular localization of polyamines in carrot protoplasts and vacuoles. *Plant Physiol* 87: 514–518.

Pitkin, J. and R. H. Davis. 1990. The genetics of polyamine synthesis in *Neurospora crassa. Arch Biochem Biophys* 278: 386–391.

Pitkin, J., M. Perriere, A. Kanehl, J. L. Ristow, and R. H. Davis. 1994. Polyamine metabolism and growth of *Neurospora* strains lacking *cis*-acting control sites in the ornithine decarboxylase gene. *Arch Biochem Biophys* 315: 153–160.

Plum, G. E. and V. A. Bloomfield. 1990. Effects of spermidine and hexaamminecobalt(III) on thymine imino proton exchange. *Biochemistry* 29: 5934–5940.

Pollard, K. J., M. L. Samuels, K. A. Crowley, J. C. Hansen, and C. L. Peterson. 1999. Functional interaction between GCN5 and polyamines: A new role for core histone acetylation. *EMBO J* 18: 5622–5633.

Polticelli, F., D. Salvi, P. Mariottini, R. Amendola, and M. Cervelli. 2012. Molecular evolution of the polyamine oxidase gene family in Metazoa. *BMC Evol Biol* 12: 1–14.

Porat, Z., G. Landau, Z. Bercovich, D. Krutauz, M. Glickman, and C. Kahana. 2008. Yeast antizyme mediates degradation of yeast ornithine decarboxylase by yeast but not by mammalian proteasome: New insights on yeast antizyme. *J Biol Chem* 283: 4528–4534.

Pottosin, I. and S. Shabala. 2014. Polyamines control cation transport across plant membranes: Implications for ion homeostasis and abiotic stress signaling. *Front Plant Sci* 5: 1–16.

Pottosin, I., A. M. Velarde-Buendia, I. Zepeda-Jazo, O. Dobrovinskaya, and S. Shabala. 2012. Synergism between polyamines and ROS in the induction of Ca (2+) and K (+) fluxes in roots. *Plant Signal Behav* 7: 1084–1087.

Poulin, R., L. Lu, B. Ackermann, P. Bey, and A. E. Pegg. 1992. Mechanism of the irreversible inactivation of mouse ornithine decarboxylase by alpha-difluoromethylornithine. Characterization of sequences at the inhibitor and coenzyme binding sites. *J Biol Chem* 267: 150–158.

Poulin, R., M. Lessard, and C. Zhao. 1995. Inorganic cation dependence of putrescine and spermidine transport in human breast cancer cells. *J Biol Chem* 270: 1695–1704.

Preeti, S. Tapas, P. Kumar, R. Madhubala, and S. Tomar. 2013. Structural insight into DFMO resistant ornithine decarboxylase from *Entamoeba histolytica*: An inkling to adaptive evolution. *PLoS One* 8: e53397.

Pritsa, T. S. and D. G. Vogiatzis. 2004. Seasonal changes in polyamines content of vegetative and reproductive olive organs in relation to floral initiation, anthesis, and fruit development. *Aus J Agri Res* 55: 1039–1046.

Pulkka, A. R., R. Ihalainen, J. Astsinki, and A. Pajunen. 1991. Structure and organization of the gene encoding rat-*S*-adenosylmethionine decarboxylase. *FEBS Lett* 291: 289–295.

Purwoko, B. S., N. Kesmayanti, S. Susanto, and M. Z. Nasution. 1998. Effect of polyamines on quality changes in papaya and mango fruits. *Acta Hort* 464: 510.

Rajam, M. V. 2012. Host induced silencing of fungal pathogen genes: An emerging strategy for disease control in crop plants. *Cell Dev Biol* 1: 6.

Rajam, M. V. and A. W. Galston. 1985. The effects of some polyamine biosynthetic inhibitors on growth and morphology of phytopathogenic fungi. *Plant Cell Physiol* 26: 683–692.

Rajam, B. and M. V. Rajam. 1996. Inhibition of polyamine biosynthesis and growth in plant pathogenic fungi *in vitro*. *Mycopathology* 133: 95–103.

Rajam, M. V., L. H. Weinstein, and A. W. Galston. 1986. Kinetic-studies on the control of the bean rust fungus (*Uromyces phaseoli* L) by an inhibitor of polyamine biosynthesis. *Plant Physiol* 82: 485–487.

Rajam, M. V., L. H. Weinstein, and A. W. Galston. 1989. Inhibition of uredospore germination. *Plant Cell Physiol* 30: 37–41.

Ramadan, N., I. Flockhart, M. Booker, N. Perrimon, and B. Mathey-Prevot. 2007. Design and implementation of high-throughput RNAi screens in cultured *Drosophila* cells. *Nat Protol* 2: 2245–2264.

Ramani, D., J. P. De Bandt, and L. Cynober. 2014. Aliphatic polyamines in physiology and diseases. *Clin Nutr* 33: 14–22.

Raspaud, E., I. Chaperon, A. Leforestier, and F. Livolant. 1999. Spermine-induced aggregation of DNA, nucleosome, and chromatin. *Biophys J* 77: 1547–1555.

Rastogi, R. and V. K. Sawhney. 1990. Polyamines and flower development in the male sterile stamenless-2 mutant of tomato (*Lycopersicon esculentum* Mill.). *Plant Physiol* 93: 446–452.

Ray, R. M., M. N. Bavaria, S. Bhattacharya, and L. R. Johnson. 2011. Activation of Dbl restores migration in polyamine-depleted intestinal epithelial cells via Rho-GTPases. *Am J Physiol Gastrointest Liver Physiol* 300: G988–G997.

Rechsteiner, M. and S. W. Rogers. 1996. PEST sequences and regulation by proteolysis. *Trends Biochem Sci* 21: 267–271.

Reis, I. A., M. P. Martinez, N. Yarlett, P. J. Johnson, F. C. Silva-Filho, and M. A. Vannier-Santos. 1999. Inhibition of polyamine synthesis arrests trichomonad growth and induces destruction of hydrogenosomes. *Antimicrob Agents Chemother* 43: 1919–1923.

Reitz, M., D. Walters, and B. Moerschbacher. 1995. Germination and appressorial formation by uredospores of *Uromyces viciae-fabae* exposed to inhibitors of polyamine biosynthesis. *Eur J Plant Pathol* 101: 573–578.

Reyna-Lopez, G. E. and J. Ruiz-Herrera. 2004. Specificity of DNA methylation changes during fungal dimorphism and its relationship to polyamines. *Curr Microbiol* 48: 118–123.

Rider, J. E., A. Hacker, C. A. Mackintosh, A. E. Pegg, P. M. Woster, and R. A. Casero, Jr. 2007. Spermine and spermidine mediate protection against oxidative damage caused by hydrogen peroxide. *Amino Acids* 33: 231–240.

Roberts, D. R., M. A. Walker, J. E. Thompson, and E. B. Dumbroff. 1984. The effects of inhibitors of polyamine and ethylene biosynthesis on senescence, ethylene production and polyamine levels in cut carnation flowers. *Plant Cell Physiol* 25: 315–322.

Roberts, S. C., J. Scott, J. E. Gasteier et al. 2002. *S*-adenosylmethionine decarboxylase from *Leishmania donovani*. Molecular, genetic, and biochemical characterization of null mutants and overproducers. *J Biol Chem* 277: 5902–5909.

Rodriguez-Garay, B., G. C. Phillips, and G. D. Kuehn. 1989. Detection of norspermidine and norspermine in *Medicago sativa* L. (Alfalfa). *Plant Physiol* 89: 525–529.

Rom, E. and C. Kahana. 1994. Polyamines regulate the expression of ornithine decarboxylase antizyme in vitro by inducing ribosomal frame-shifting. *Proc Natl Acad Sci U S A* 91: 3959–3963.

Rosenblum, M. G., M. J. Keating, B. S. Yap, and T. L. Loo. 1981. Pharmacokinetics of ([14C]methylglyoxal-bis-guanylhydrazone) in patients with leukemia. *Cancer Res* 41: 1748–1750.

Rosi, S., R. Ferguson, K. Fishman, A. Allen, J. Raber, and J. R. Fike. 2012. The polyamine inhibitor alpha-difluoromethylornithine modulates hippocampus dependent function after single and combined injuries. *PLoS One* 7: 1–10.

Rowland, H. D., D. R. Morris, and P. Coffino. 1992. Sequestered end products and enzyme regulation: The case of ornithine decarboxylase. *Microbiol Rev* 56: 280–290.

Ruan, H., L. M. Shantz, A. E. Pegg, and D. R. Morris. 1996. The upstream open reading frame of the mRNA encoding *S*-adenosylmethionine decarboxylase is a polyamine-responsive translational control element. *J Biol Chem* 271: 29576–29582.

Ruhl, M., M. Himmelspach, G. M. Bahr et al. 1993. Eukaryotic initiation factor 5A is a cellular target of the human immunodeficiency virus type 1 Rev activation domain mediating trans-activation. *J Cell Biol* 123: 1309–1320.

Ruiz-Chica, J., M. A. Medina, F. Sanchez-Jimenez, and F. J. Ramirez. 2003. Raman spectroscopy study of the interaction between biogenic polyamines and an alternating AT oligodeoxyribonucleotide. *Biochim Biophys Acta* 1628: 11–21.

Ruiz-Herrera, J. 1984. El dimorfismo de los hongos como modelo de diferenciación bioquímica. In *Caminos en la Biología Fundamental*, eds. J. Martuscelli, R. Palacios de la Lama, and G. Soberón Acevedo, 285–303. México: Universidad Autónoma de México.

Ruiz-Herrera, J. 1994. Polyamines, DNA methylation, and fungal differentiation. *Crit Rev Microbiol* 20: 143–150.

Ruiz-Herrera, J. and C. Calvo-Mendez. 1987. Effect of ornithine decarboxylase inhibitors on the germination of sporangiospores of Mucorales. *Exp Mycol* 11: 287–296.

Ruiz-Herrera, J. and T. Hard. 1997. Dimorphism in *Mucor rouxii* (Zygomycetes). *Publications Wissenschatlichen Film Biologie* 23: 17–32.

Ruiz-Herrera, J., A. Ruiz, and E. Lopez-Romero. 1983. Isolation and biochemical analysis of *Mucor bacilliformis* monomorphic mutants. *J Bacteriol* 156: 262–272.

Ruiz-Herrera, J., R. Ruiz-Medrano, and A. Domínguez. 1995. Selective inhibition of cytosine–DNA methylases by polyamines. *FEBS Lett* 357: 192–196.

Russell, D. H. and S. H. Snyder. 1969. Amine synthesis in regenerating rat liver: Extremely rapid turnover of ornithine decarboxylase. *Mol Pharmacol* 5: 253–262.

Russell, D. H., H. R. Giles, C. D. Christian, and J. L. Campebell. 1978. Polyamines in amniotic fluid, plasma, and urine during normal pregnancy. *Am J Obstet Gynecol* 15: 642–652.

Rutto, K. L., T. Ishii, L. S. Wamocho, and M. Murakami. 1999. The effect of polyamines and propamocarb on the growth and yield of tomato (*Lycopersicon esculentum* Mill.) plants grown in a low P tropical soil inoculated with a vesicular-arbuscular mycorrhizal fungus. *Bull Fac Educ Ehime Univ Nat Sci* 19: 33–37.

Sadasivan, S. K., B. Vasamsetti, J. Singh, V. V. Marikunte, A. M. Oommen, M. R. Jagannath, and R. R. Pralhada. 2014. Exogenous administration of spermine improves glucose utilization and decreases bodyweight in mice. *Eur J Pharmacol* 729: 94–99.

Saftner, R. A. and W. S. Conway. 1997. Effects of some polyamine biosynthesis inhibitors and calcium chloride on in vitro growth and decay development in apples caused by *Botrytis cinerea* and *Penicillium expansum*. *J Am Soc Hort Sci* 122: 380–385.

Sagor, G. H., T. Berberich, Y. Takahashi, M. Niitsu, and T. Kusano. 2013. The polyamine spermine protects *Arabidopsis* from heat stress-induced damage by increasing expression of heat shock-related genes. *Transgenic Res* 22: 595–605.

Saini, P., D. E. Eyler, R. Green, and T. E. Dever. 2009. Hypusine-containing protein eIF5A promotes translation elongation. *Nature* 459: 118–121.

Saito, H. 1985. Alterations of polyamines in body fluids during pregnancy in rats. *Nihon Sanka Fujinka Gakkai Zasshi* 37: 293–300.

Sakai, T., C. Hori, K. Kano, and T. Oka. 1979. Purification and characterization of *S*-adenosyl-L-methionine decarboxylase from mouse mammary gland and liver. *Biochemistry* 18: 5541–5548.

Sakata, S. F., L. L. Shelly, S. Ruppert, G. Schutz, and J. Y. Chou. 1993. Cloning and expression of murine *S*-adenosylmethionine synthetase. *J Biol Chem* 268: 13978–13986.

Sakurada, T. 1962. Effect of spermine on alcoholic fermentation of bakers' yeast. *J Biochem* 52: 245–249.

Salas-Marco, J. and D. M. Bedwell. 2004. GTP hydrolisis by eRF3 facilitates stop codon decoding during eukaryotic translation termination. *Mol Cell Biol* 24: 7769–7778.

Salunke, D. B., B. G. Hazra, and V. S. Pore. 2003. Bile acid–polyamine conjugates as synthetic ionophores. *ARKAT USA* ix: 115–125.

Samal, K., P. Zhao, A. Kendzicky et al. 2013. AMXT-1501, a novel polyamine transport inhibitor, synergizes with DFMO in inhibiting neuroblastoma cell proliferation by targeting both ornithine decarboxylase and polyamine transport. *Int J Cancer* 133: 1323–1334.

San-Blas, G., F. Sorais, F. San-Blas, and J. Ruiz-Herrera. 1996. Ornithine decarboxylase in *Paracoccidioides brasiliensis*. *Arch Microbiol* 165: 311–316.

San-Blas, G., F. San-Blas, F. Sorais, B. Moreno, and J. Ruiz-Herrera. 1997. Polyamines in growth and dimorphism of *Paracoccidioides brasiliensis*. *Arch Microbiol* 166: 411–413.

Sanches, M., C. A. O. Dias, L. H. Aponi, S. R. Valentini, and B. Guimaraes. 2008. Crystal structure of the full length eIF5A from *Saccharomyces cerevisiae*. RCSB-Protein Data Bank. Available at http://www.rcsb.org/.

Sandler, S., K. Bendtzen, D. L. Eizirik, A. Sjoholm, and N. Welsh. 1989. Decreased cell replication and polyamine content in insulin-producing cells after exposure to human interleukin 1 beta. *Immunol Lett* 22: 267–272.

Sannazzaro, A. I., C. L. Alvarez, A. B. Menendez, F. L. Pieckenstain, E. O. Alberto, and O. A. Ruiz. 2004. Ornithine and arginine decarboxylase activities and effect of some polyamine biosynthesis inhibitors on *Gigaspora rosea* germinating spores. *FEMS Microbiol Lett* 230: 115–121.

Sasaki, K., M. R. Abid, and M. Miyazaki. 1996. Deoxyhypusine synthase gene is essential for cell viability in the yeast *Saccharomyces cerevisiae*. *FEBS Lett* 384: 151–154.

Savarese, T. M., L. Y. Ghoda, D. L. Dexter, and R. E. Parks, Jr. 1983. Conversion of 5′-deoxy-5′-methylthioadenosine and 5′-deoxy-5′-methylthioinosine to methionine in cultured human leukemic cells. *Cancer Res* 43: 4699–4702.

Sawhney, R. K., N. S. Shekhawat, and A. W. Galston. 1985. Polyamine levels as related to growth, differentiation and senescence in protoplast-derived cultures of *Vignia aconitifolia* and *Avena sativa*. *Plant Growth Regul* 3: 329–337.

Scarpa, S., M. Lucarelli, F. Palitti, D. Carotti, and R. Strom. 1996. Simultaneous myogenin expression and overall DNA hypomethylation promote in vitro myoblast differentiation. *Cell Growth Differ* 7: 1051–1058.

Schipper, R. G. and A. A. Verhofstad. 2002. Distribution patterns of ornithine decarboxylase in cells and tissues: Facts, problems, and postulates. *J Histochem Cytochem* 50: 1143–1160.

Schmid, N. and J. P. Behr. 1991. Location of spermine and other polyamines on DNA as revealed by photoaffinity cleavage with polyaminobenzenediazonium salts. *Biochemistry* 30: 4357–4361.

Schneider, B. L. and L. Reitzer. 2012. Pathway and enzyme redundancy in putrescine catabolism in *Escherichia coli*. *J Bacteriol* 194: 4080–4088.

Schneider, B. L., V. J. Hernandez, and L. Reitzer. 2013. Putrescine catabolism is a metabolic response to several stresses in *Escherichia coli*. *Mol Microbiol* 88: 537–550.

Schnier, J., H. G. Schwelberger, Z. Smit-McBride, H. A. Kang, and J. W. Hershey. 1991. Translation initiation factor 5A and its hypusine modification are essential for cell viability in the yeast *Saccharomyces cerevisiae*. *Mol Cell Biol* 11: 3105–3114.

Schrader, R., C. Young, D. Kozian, R. Hoffmann, and F. Lottspeich. 2006. Temperature-sensitive eIF5A mutant accumulates transcripts targeted to the nonsense-mediated decay pathway. *J Biol Chem* 281: 35336–35346.

Schuber, F. 1989. Influence of polyamines on membrane functions. *Biochem J* 260: 1–10.

Schubler, A., D. Schwarzott, and C. Walker. 2001. A new fungal phylum, the *Glomeromycota*: Phylogeny and evolution. *Mycol Res* 12: 1413–1421.

Schuster, I. and R. Bernhardt. 2011. Interaction of natural polyamines with mammalian proteins. *Biol Concepts* 2: 79–94.

Schwartz, M. F., S. J. Lee, J. K. Duong, S. Eminaga, and D. F. Stern. 2003. FHA domain-mediated DNA checkpoint regulation of Rad53. *Cell Cycle* 2: 384–396.

Schwechheimer, C. and X. W. Deng. 2001. COP9 signalosome revisited: A novel mediator of protein degradation. *Trends Cell Biol* 11: 420–426.

Scorcioni, F., A. Corti, P. Davalli, S. Astancolle, and P. S. Bettuzzi. 2001. Manipulation of the expression of regulatory genes of polyamine metabolism results in specific alterations of the cell-cycle progression. *Biochem J* 354: 217–223.

Scuoppo, C., C. Miething, L. Lindqvist et al. 2012. A tumour suppressor network relying on the polyamine–hypusine axis. *Nature* 487: 244–248.

Sebela, M., A. Radova, R. Angelini, P. Tavladoraki, I. I. Frebort, and P. Pec. 2001. FAD-containing polyamine oxidases: A timely challenge for researchers in biochemistry and physiology of plants. *Plant Sci* 160: 197–207.

Seely, J. E., H. Pösö, and A. E. Pegg. 1982. Purification of ornithine decarboxylase from kidneys of androgen-treated mice. *Biochemistry* 21: 3394–3399.

Seif El-Yazal, M. A. and M. M. Rady. 2012. Changes in nitrogen and polyamines during breaking bud dormancy in "Anna" apple trees with foliar applications of some compounds. *Sci Hort* 136: 75–80.

Seiler, N. 1987. Functions of polyamine acetylation. *Can J Physiol Pharmacol* 65: 2024–2035.

Seiler, N. 1995. Polyamine oxidase, properties and functions. *Prog Brain Res* 106: 333–344.

Seiler, N. 2004. Catabolism of polyamines. *Amino Acids* 26: 217–233.

Seiler, N. and B. Eichentopf. 1975. 4-aminobutyrate in mammalian putrescine catabolism. *Biochem J* 152: 201–210.

Seiler, N. and F. Raul. 2007. Polyamines and the intestinal tract. *Crit Rev Clin Lab Sci* 44: 365–411.

Seiler, N., S. Sarhan, and B. F. Roth-Schechter. 1981. Developmental changes of the GABA and polyamine systems in isolated neurons in cell culture. *Dev Neurosci* 4: 181–187.

Shah, N., T. Antony, S. Haddad, P. Amenta, A. Shirahata, T. J. Thomas, and T. Thomas. 1999a. Antitumor effects of bis(ethyl) polyamine analogs on mammary tumor development in FVB/NTgN (MMTVneu) transgenic mice. *Cancer Lett* 146: 15–23.

Shah, N., T. Thomas, A. Shirahata, L. H. Sigal, and T. J. Thomas. 1999b. Activation of nuclear factor kappa B by polyamines in breast cancer cells. *Biochem* 38: 14763–14774.

Shantz, L. M. and A. E. Pegg. 1999. Translational regulation of ornithine decarboxylase and other enzymes of the polyamine pathway. *Int J Biochem Cell Biol* 31: 107–122.

Shantz, L. M., I. Holm, O. A. Janne, and A. E. Pegg. 1992. Regulation of *S*-adenosylmethionine decarboxylase activity by alterations in the intracellular polyamine content. *Biochem J* 288: 511–518.

Shapira, R., A. Altman, Y. Henis, and I. Chet. 1989. Polyamines and ornithine decarboxylase activity during growth and differentiation in *Sclerotium rolfsii*. *J Gen Microbiol* 135: 1361–1367.

Shelp, B. J., G. G. Bozzo, C. P. Trobacher, A. Zarei, K. L. Deyman, and C. J. Brikis. 2012. Hypothesis/review: Contribution of putrescine to 4-aminobutyrate (GABA) production in response to abiotic stress. *Plant Sci* 193–194: 130–135.

Shevyakova, N. I., L. I. Musatenko, L. A. Stetsenko, N. P. Vedenicheva, L. P. Voitenko, K. M. Sytnik, and Vl. V. Kuznetsov. 2013. Effect of abscisic acid on the contents of polyamines and proline in common bean plants under salt stress. *Russ J Plant Physiol* 60: 200–211.

Shi, H. and Z. Chan. 2014. Improvement of plant abiotic stress tolerance through modulation of the polyamine pathway. *J Integr Plant Biol* 56: 114–121.

Shimogori, T., K. Kashiwagi, and K. Igarashi. 1996. Spermidine regulation of protein synthesis at the level of initiation complex formation of Met-tRNAi, mRNA and ribosomes. *Biochem Biophys Res Commun* 223: 544–548.

Shirahata, A. and A. E. Pegg. 1985. Regulation of *S*-adenosylmethionine decarboxylase activity in rat liver and prostate. *J Biol Chem* 260: 9583–9588.

Shirahata, A. and A. E. Pegg. 1986. Increased content of mRNA for a precursor of *S*-adenosylmethionine decarboxylase in rat prostate after treatment with 2-difluoromethylornithine. *J Biol Chem* 261: 13833–13837.

Shirahata, A., N. Takahashi, T. Beppu, H. Hosoda, and K. Samejima. 1993. Effects of inhibitors of spermidine synthase and spermine synthase on polyamine synthesis in rat tissues. *Biochem Pharmacol* 45: 1897–1903.

Shiraki, K., M. Kudou, Y. Aso, and M. Takagi. 2003. Dissolution of protein aggregation by small amine compounds. *Sci Technol Adv Mat* 4: 55–59.

Shoeb, F., J. S. Yadav, S. Bajaj, and M. V. Rajam. 2001. Polyamines as biomarkers for plant regeneration capacity: Improvement of regeneration by modulation of polyamine metabolism in different genotypes of indica rice. *Plant Sci* 160: 1229–1235.

Shore, L. J., A. P. Soler, and S. K. Gilmour. 1997. Ornithine decarboxylase expression leads to translocation and activation of protein kinase CK2 in vivo. *J Biol Chem* 272: 12536–12543.

Sichhar, Y. and B. Dräger. 2013. Immunolocalisation of spermidine synthase in *Solanum tuberosum*. *Phytochemistry* 91: 117–121.

Siibak, T. and J. Remme. 2010. Subribosomal particle analysis reveals the stages of bacterial ribosome assembly at which rRNA nucleotides are modified. *RNA* 16: 2023–2032.

Sindhu, R. K. and S. S. Cohen. 1984. Subcellular localization of spermidine synthase in the protoplasts of Chinese cabbage leaves. *Plant Physiol* 76: 219–223.

Slocum, R. D. 1991. Tissue and subcellular localization of polyamines and enzymes of polyamine metabolism. In *Biochemistry and Physiology of Polyamines in Plants*, eds. R. D. Slocum and H. E. Flores, 93–103. Boca Raton, FL: CRC Press.

Slocum, R. D. and A. W. Galston. 1985. *In vivo* inhibition of polyamine biosynthesis and growth in tobacco ovary tissues. *Plant Cell Physiol* 26: 1519–1526.

Slocum, R. D., R. Kaur-Sawhney, and A. W. Galston. 1984. The physiology and biochemistry of polyamines in plants. *Arch Biochem Biophys* 235: 283–303.

Smith, S. E. and D. J. Read. 1997. *Mycorrhizal Symbiosis*. New York: Academic Press.

Smith, T. A., J. H. A. Barker, and M. Jung. 1990. Growth-inhibition of *Botrytis cinerea* by compounds interfering with polyamine metabolism. *J Gen Microbiol* 136: 985–992.

Smith, T. A., J. H. A. Barker, and W. J. Owen. 1992. Insensitivity of *Septoria tritici* and *Ustilago maydis* to inhibitors of ornithine decarboxylase. *Mycol Res* 96: 395–400.

Sobieszczuk-Nowicka, E. and J. Legocka. 2014. Plastid-associated polyamines: Their role in differentiation, structure, functioning, stress response and senescence. *Plant Biol* 16: 297–305.

Soda, K., Y. Kano, F. Chiba, K. Koizumi, and Y. Miyaki. 2013. Increased polyamine intake inhibits age-associated alteration in global DNA methylation and 1,2-Dimethylhydrazine-induced tumorigenesis. *PLoS One* 8: e64357.

Soulet, D., L. Covassin, M. Kaouass, R. Charest-Gaudreault, M. Audette, and R. Polulin. 2002. Role of endocitosis in the internalization of spermidine-C2-BODIPY, a highly fluorescent probe of polyamine transport. *Biochem J* 367: 347–357.

Soulet, D., B. Gagnon, S. Rivest, M. Audette, and R. Poulin. 2004. A fluorescent probe of polyamine transport accumulates into intracellular acidic vesicles via a two-step mechanism. *J Biol Chem* 279: 49355–49366.

Sparapani, M., M. Virgili, G. Bardi, M. Treqnago, B. Monti, M. Bentivoqli, and A. Contestabile. 1998. Ornithine decarboxylase activity during development of cerebellar granule neurons. *J Neurochem* 71: 1898–1904.

Spotheim-Maurizot, M., S. Ruiz, R. Sabattier, and M. Charlier. 1995. Radioprotection of DNA by polyamines. *Int J Radiat Biol* 68: 571–577.

Srivastava, A., S. H. Chung, T. Fatima, T. Datsenka, A. K. Handa, and A. K. Mattoo. 2007. Polyamines as anabolic growth regulators revealed by transcriptome analysis and metabolite profiles of tomato fruits engineered to accumulate spermidine and spermine. *Plant Biotechnol* 24: 57–70.

Stanley, B. A. and A. E. Pegg. 1991. Amino acid residues necessary for putrescine stimulation of human *S*-adenosylmethionine decarboxylase proenzyme processing and catalytic activity. *J Biol Chem* 266: 18502–18506.

Stanley, B. A., A. E. Pegg, and I. Holm. 1989. Site of pyruvate formation and processing of mammalian *S*-adenosylmethionine decarboxylase proenzyme. *J Biol Chem* 264: 21073–21079.

Stefanis, L. 2012. Alpha-Synuclein in Parkinson's disease. *Cold Spring Harb Perspect Med* 2: a009399.

Steitz, T. A. 1990. Structural studies of protein–nucleic acid interaction: The sources of sequence-specific binding. *Q Rev Biophys* 23: 205–280.

Stevens, L. and M. Winther. 1979. Spermine, spermidine and putrescine in fungal development. *Adv Microb Physiol* 19: 63–148.

Stevens, L., I. M. McKinnon, and M. Winther. 1977. The effects of 1,4-di-aminobutanone on polyamine synthesis in *Aspergillus nidulans*. *FEBS Lett* 75: 180–182.

Stillway, L. W. and T. Walle. 1977. Identification of the unusual polyamines 3,3′-diaminodipropylamine and N,N′-bis(3-aminoproply)-1,3-propanediamine in the white shrimp *Penaeus setiferus*. *Biochem Biophys Res Commun* 77: 1103–1107.

Stuehr, D. J. 2004. Enzymes of the L-arginine to nitric oxide pathway. *J Nutr* 134: 2748S–2751S.

Subhi, A. L., P. Diegelman, C. W. Porter, B. Tang, Z. J. Lu, G. D. Markham, and W. D. Kruger. 2003. Methylthioadenosine phosphorylase regulates ornithine decarboxylase by production of downstream metabolites. *J Biol Chem* 278: 49868–49873.

Sugiyama, S., Y. Matsuo, K. Maenaka, D. G. Vassylyev, and M. Matsushima. 1996. The 1.8-A X-ray structure of the *Escherichia coli* PotD protein complexed with spermidine and the mechanism of polyamine binding. *Protein Sci* 5: 1984–1990.

Suh, J. W., S. H. Lee, B. C. Chung, and J. Park. 1997. Urinary polyamine evaluation for effective diagnosis of various cancers. *J Chromatogr* 688: 179–186.

Sunkara, P. S., P. N. Rao, K. Nishioka, and B. R. Brinkley. 1979. Role of polyamines in cytokinesis of mammalian cells. *Exp Cell Res* 119: 63–68.

Supattapone, S., H. Willie, L. Uyechi et al. 2001. Branched polyamines cure prion infected neuroblastoma cells. *J Virol* 75: 3453–3461.

Suzuki, T., Y. He, K. Kashiwagi, Y. Murakami, S. Hayashi, and K. Igarashi. 1994. Antizyme protects against abnormal accumulation and toxicity of polyamines in ornithine decarboxylase-overproducing cells. *Proc Natl Acad Sci U S A* 91: 8930–8934.

Suzuki, T., Y. K. Kim, H. Yoshioka, and Y. Iwahashi. 2013. Regulation of metabolic products and gene expression in *Fusarium asiaticum* by agmatine addition. *Mycotoxin Res* 29: 103–111.

Szumanski, M. B. and S. M. Boyle. 1990. Analysis and sequence of the speB gene encoding agmatine ureohydrolase, a putrescine biosynthetic enzyme in *Escherichia coli*. *J Bacteriol* 172: 538–547.

Tabor, C. W. and H. Tabor. 1984. Polyamines. *Annu Rev Biochem* 53: 749–790.

Tabor, C. W. and H. Tabor. 1987. The speEspeD operon of *Escherichia coli*. Formation and processing of a proenzyme form of *S*-adenosylmethionine decarboxylase. *J Biol Chem* 262: 16037–16040.

Tadolini, B. 1988. Polyamine inhibition of lipoperoxidation. The influence of polyamines on iron oxidation in the presence of compounds mimicking phospholipid polar heads. *Biochem J* 249: 33–36.

Takahashi, T. and J. I. Kakehi. 2010. Polyamines: Ubiquitous polycations with unique roles in growth and stress responses. *Ann Bot* 105: 1–6.

Takano, A., J. Kakehi, and T. Takahashi. 2012. Thermospermine is not a minor polyamine in the plant kingdom. *Plant Cell Physiol* 53: 606–616.

Tamilarasan, S. and M. V. Rajam. 2013. Engineering crop plants for nematode resistance through host-derived RNA interference. *Cell Dev Biol* 2: 114.

Tassoni, A., S. Fornale, and N. Bagni. 2003. Putative ornithine decarboxylase activity in *Arabidopsis thaliana*: Inhibition and intracellular localisation. *Plant Physiol Biochem* 41: 871–875.

Tavladoraki, P., M. E. Schinina, F. Cecconi et al. 1998. Maize polyamine oxidase: Primary structure from protein and cDNA sequencing. *FEBS Lett* 426: 62–66.

Teng, Y. B., Y. X. He, Y. L. Jiang, Y. X. Chen, and C. Z. Zhou. 2009. Crystal structure of eukaryotic translation initiation factor eIF5A2 from *Arabidopsis thaliana*. *Proteins* 77: 736–740.

Terakado, J., T. Yoneyama, and S. Fujihara. 2006. Shoot-applied polyamines suppress nodule formation in soybean (*Glycine max*). *J Plant Physiol* 163: 497–505.

Teti, D., M. Visalli, and H. McNair. 2002. Analysis of polyamines as markers of (patho)physiological conditions. *J Chromatogr* 781: 107–149.

Thomas, T. and D. T. Kiang. 1988. Modulation of the binding of progesterone receptor to DNA by polyamines. *Cancer Res* 48: 1217–1222.

Thomas, T. J. and R. P. Messner. 1988. Structural specificity of polyamines in left-handed Z-DNA formation. Immunological and spectroscopic studies. *J Mol Biol* 201: 463–467.

Thomas, T. J., N. L. Meryhew, and R. P. Messner. 1990. Enhanced binding of lupus sera to the polyamine-induced left-handed Z-DNA form of polynucleotides. *Arthritis Rheum* 33: 356–365.

Thomas, T., M. A. Gallo, C. M. Klinge, and T. J. Thomas. 1995. Polyamine-mediated conformational perturbations in DNA alter the binding of estrogen receptor to poly(dG-m5dC). poly(dG-m5dC) and a plasmid containing the estrogen response element. *J Steroid Biochem Mol Biol* 54: 89–99.

Tiburcio, A. F., E. Cohen, S. M. Arad, E. Birnbaum, and Y. Mizrahi. 1985. Polyamine metabolism. In *Intermediary Nitrogen Metabolism: The Biochemistry of Plants*, ed. B. J. Mifflin, 283–325. New York: Academic Press.

Todd, B. A., V. A. Parsegian, A. Shirahata, T. J. Thomas, and D. C. Rau. 2008. Attractive forces between cation condensed DNA double helices. *Biophys J* 94: 4775–4782.

Todorova, D. I., V. Sergiev, E. Alexieva, A. Karanov, A. Smith, and M. Hall. 2007. Polyamine content in *Arabidopsis thaliana* (L.) Heynh during recovery after low and high temperature treatments. *Plant Growth Regul* 51: 185–191.

Tolbert, W. D., D. E. Graham, R. H. White, and S. E. Ealick. 2003. Pirovoyl-dependent arginine decarboxylase from *Methanococcus jannaschii*: Crystal structures of the self-cleaved and S53A proenzyme forms. *Structure* 11: 285–294.

Tong, Y., I. Park, B. S. Hong, L. Nedyalkova, W. Tempel, and H. W. Park. 2009. Crystal structure of human eIF5A1: Insight into functional similarity of human eIF5A1 and eIF5A2. *Proteins* 75: 1040–1045.

Toninello, A., L. D. Via, V. Di Noto, and M. Mancon. 1999. The effects of methylglyoxal-bis(guanylhydrazone) on spermine binding and transport in liver mitochondria. *Biochem Pharmacol* 58: 1899–1906.

Toninello, A., M. Salvi, and B. Mondovi. 2004. Interaction of biologically active amines with mitochondria and their role in the mitochondrial-mediated pathway of apoptosis. *Curr Med Chem* 11: 2349–2374.

Torres-Guzman, J. C., B. Xoconostle-Cazares, L. Guevara-Olvera, L. Ortiz, G. San-Blas, A. Dominguez, and J. Ruiz-Herrera. 1996. Comparison of fungal ornithine decarboxylases. *Curr Microbiol* 33: 390–392.

Torrigiani, P., D. Serafini-Fracassini, S. Biondi, and N. Bagni. 1986. Evidence for the subcellular localization of polyamines and their biosynthetic enzymes in plant cells. *J Plant Physiol* 124: 23–29.

Torrigiani, P., A. M. Bregoli, V. Ziosi et al. 2004. Pre-harvest polyamine and aminoethoxyvinylglycine (AVG) applications modulate fruit ripening in Stark Red Gold nectarines (Prunus persica L. Batsch). *Postharvest Biol Technol* 33: 293–308.

Tran, M. K., C. J. Schultz, and U. Baumann. 2008. Conserved upstream open reading frames in higher plants. *BMC Genomics* 9: 361.

Trione, E. J., O. Stockwell, and H. A. Austin. 1988. The effects of polyamines on the growth and development of the wheat bunt fungi. *Bot Gaz* 149: 173–178.

Trull, M. C., B. L. Holaway, and R. L. Malmberg. 1992. Development of stigmatoid anthers in a tobacco mutant: Implications for regulation of stigma differentiation. *Can J Bot* 70: 2339–2346.

Tsaniklidis, G., C. Delis, N. Nikolaudakis, P. Katinakis, and G. Aivalakis. 2014. Low temperatures storage effects the ascorbic acid metabolism of cherry tomato fruits. *Plant Physiol Biochem* 84: 149–157.

Tsuji, T., S. Usui, T. Aida et al. 2001. Induction of epithelial differentiation and DNA demethylation in hamster malignant oral keratinocyte by ornithine decarboxylase antizyme. *Oncogene* 20: 24–33.

Tyagi, A. K., C. W. Tabor, and H. Tabor. 1981. Ornithine decarboxylase from *Saccharomyces cerevisiae*. Purification, properties, and regulation of activity. *J Biol Chem* 256: 12156–12163.

Tylichova, M., D. Kopecny, S. Morera, P. Briozzo, R. Lenobel, J. Snegaroff, and M. Sebela. 2010. Structural and functional characterization of plant aminoaldehyde dehydrogenase from *Pisum sativum* with a broad specificity for natural and synthetic aminoaldehydes. *J Mol Biol* 396: 870–882.

Ueda, M., N. Kanayama, N. Kamasawa, M. Osumi, and A. Tanaka. 2003. Up-regulation of the peroxisomal beta-oxidation system occurs in butyrate-grown *Candida tropicalis* following disruption of the gene encoding peroxisomal 3-ketoacyl-CoA thiolase. *Biochim Biophys Acta* 1631: 160–168.

Urakov, V. N., I. A. Valouev, N. V. Kochneva-Pervukhova et al. 2006. *N*-terminal region of *Saccharomyces cerevisiae* eRF3 is essential for the functioning of the eRF1/eRF3 complex beyond translation termination. *BMC Mol Biol* 11: 34.

Urano, K., Y. Yoshiba, T. Nanjo et al. 2003. Characterization of *Arabidopsis* genes involved in biosynthesis of polyamines in abiotic stress responses and developmental stages. *Plant Cell Environ* 26: 1917–1926.

Urano, K., T. Hobo, and K. Shinizaki. 2005. *Arabidopsis* ADC genes involved in polyamine biosynthesis are essential for seed development. *FEBS Lett* 579: 1557–1564.

Valdes-Santiago, L. and J. Ruiz-Herrera. 2014. Stress and polyamine metabolism in fungi. *Front Chem* 1: 1–10.

Valdes-Santiago, L., J. A. Cervantes-Chavez, and J. Ruiz-Herrera. 2009. *Ustilago maydis* spermidine synthase is encoded by a chimeric gene, required for morphogenesis, and indispensable for survival in the host. *FEMS Yeast Res* 9: 923–935.

Valdes-Santiago, L., D. Guzman-de-Pena, and J. Ruiz-Herrera. 2010. Life without putrescine: Disruption of the gene-encoding polyamine oxidase in *Ustilago maydis odc* mutants. *FEMS Yeast Res* 10: 928–940.

Valdes-Santiago, L., J. A. Cervantes-Chavez, C. G. Leon-Ramirez, and J. Ruiz-Herrera. 2012a. Polyamine metabolism in fungi with emphasis on phytopathogenic species. *J Amino Acids* 2012: 1–13.

Valdes-Santiago, L., J. A. Cervantes-Chavez, R. Winkler, C. G. Leon-Ramirez, and J. Ruiz-Herrera. 2012b. Phenotypic comparison of samdc and spe mutants reveals complex relationships of polyamine metabolism in *Ustilago maydis*. *Microbiology* 158: 674–684.

Valdes-Santiago, L., J. A. Cervantes-Chavez, C. G. Leon-Ramirez, and J. Ruiz-Herrera. 2012c. Polyamine metabolism in fungi with emphasis on phytopathogenic species. *J Amino Acids* 2012: 837932.

Valero, D., D. Martínez-Romero, and M. Serrano. 2002. The role of polyamines in the improvement of the shelf life of fruit. *Trends Food Sci Technol* 13: 228–234.

van Dam, L., N. Korolev, and L. Nordenskiold. 2002. Polyamine–nucleic acid interactions and the effects on structure in oriented DNA fibers. *Nucleic Acids Res* 30: 419–428.

Van den Munckhof, R. J., M. Denyn, W. Tigchelaar-Gutter, R. G. Schipper, A. A. Verhofstad, C. J. Van Noorden, and W. M. Frederiks. 1995. *In situ* substrate specificity and ultrastructural localization of polyamine oxidase activity in unfixed rat tissues. *J Histochem Cytochem* 43: 1155–1162.

van der Klei, I. J. and M. Veenhuis. 2006. Yeast and filamentous fungi as model organisms in microbody research. *Biochim Biophys Acta* 1763: 1364–1373.

Van Nieuwenhove, S., P. J. Schechter, J. Declercq, G. Bone, J. Burke, and A. Sjoerdsma. 1985. Treatment of gambiense sleeping sickness in the Sudan with oral DFMO (DL-alpha-difluoromethylornithine), and inhibitor of ornithine decarboxylase; first field trial. *Trans R Soc Trop Med Hyg* 79: 692–698.

van Steeg, H., C. T. van Oostrom, H. M. Hodemaekers, and C. F. van Kreyl. 1990. Cloning and functional analysis of the rat ornithine decarboxylase-encoding gene. *Gene* 93: 249–256.

Vannier-Santos, M. A., D. Menezes, M. F. Oliveira, and F. G. de Mello. 2008. The putrescine analogue 1,4-diamino-2-butanone affects polyamine synthesis, transport, ultrastructure and intracellular survival in *Leishmania amanozonesis*. *Microbiology* 154: 3104–3111.

Venkiteswaran, S., T. Thomas, and T. J. Thomas. 2006. Role of polyamines in regulation of sequence-specific DNA binding activity. In *Polyamine Cell Signaling*, eds. J. Y. Wang and R. A. Casero, 91–122. New Jersey: Humana Press.

Vida, N., H. Svobodova, L. Rarova, P. Drasar, D. Saman, J. Cvacka, and Z. Wimmer. 2012. Polyamine conjugates of stigmasterol. *Steroids* 77: 1212–1218.

Viotti, A., N. Bagni, E. Sturani, and F. A. M. Alberghina. 1971. Magnesium and polyamine levels in *Neurospora crassa* mycelia. *Biochim Biophys Acta* 244: 329–337.

Visvanathan, A., K. Ahmed, L. Even-Faitelson, D. Lleres, D. P. Bazett-Jones, and A. I. Lamond. 2013. Modulation of higher order chromatin conformation in mammalian cell nuclei can be mediated by polyamines and divalent cations. *PLoS One* 8: 1–16.

Voigt, J., B. Deinert, and P. Bohley. 1999. Subcellular localization and light–dark control of ornithine decarboxylase in the unicellular green alga *Chlamydomonas reinhardtii*. *Physiol Platarum* 108: 353–360.

von Arnim, A. G. 2003. On again–off again: COP9 signalosome turns the key on protein degradation. *Curr Opin Plant Biol* 6: 520–529.

von Besser, H., G. Niemann, B. Domdey, and R. D. Walter. 1995. Molecular cloning and characterization of ornithine decarboxylase cDNA of the nematode *Panegrellus redivivus*. *Biochem J* 308: 635–640.

Von Hoff, D. D. 1994. MGBG: Teaching an old drug new tricks. *Ann Oncol* 5: 487–493.

Vujcic, S., P. Liang, P. Diegelman, D. L. Kramer, and C. W. Porter. 2003. Genomic identification and biochemical characterization of the mammalian polyamine oxidase involved in polyamine back-conversion. *Biochem J* 370: 19–28.

Wal, M. and B. F. Pugh. 2012. Genome-wide mapping of nucleosome positions in yeast using high-resolution MNase ChIP-Seq. *Methods Enzymol* 513: 233–250.

Wallace, H. M. 1998. Polyamines: Specific metabolic regulators or multifunctional polycations? *Biochem Soc Trans* 26: 569–571.

Wallace, H. M. and A. V. Fraser. 2003. Polyamine analogues as anticancer drugs. *Biochem Soc Trans* 31: 393–396.

Wallace, H. M. and A. V. Fraser. 2004. Inhibitors of polyamine metabolism: Review article. *Amino Acids* 26: 353–365.

Wallace, H. M. and K. Niiranen. 2007. Polyamine analogues—An update. *Amino Acids* 33: 261–265.

Walters, D. R. 1986. The effects of a polyamine biosynthesis inhibitor on infection of *Vicia faba* L. by the rust fungus, *Uromyces viciae-fabae* (Pers.) Schroet. *New Phytol* 104: 613–619.

Walters, D. R. 1995. Inhibition of polyamine biosynthesis in fungi. *Mycol Res* 99: 129–139.

Walters, D. R. 1997. The putrescine analogue (E)-1,4-diaminobut-2-ene reduces DNA methylation in the plant pathogenic fungus *Pyrenophora avenae*. *FEMS Microbiol Lett* 154: 215–218.

Walters, D. R. 2000. Polyamines in plant-microbe interactions. *Physiol Mol Plant Pathol* 57: 137–146.

Walters, D. R., T. Cowley, and A. McPherson. 1997. Polyamine metabolism in the thermotolerant mesophilic fungus *Aspergillus fumigatus*. *FEMS Microbiol Lett* 153: 433–437.

Walters, R. L., G. L. Newton, P. L. Olive, and R. C. Fahey. 1999. Radioprotection if human cell nuclear DNA by polyamines: Radiosensitivity of chromatin is influenced by tightly bound spermine. *Radiat Res* 151: 354–362.

Wang, J. Y. and L. R. Johnson. 1994. Expression of protooncogenes c-fos and c-myc in healing of gastric mucosal stress ulcers. *Am J Physiol* 266: G878–G886.

Wang, J. Y., S. A. McCormack, M. J. Viar, H. Wang, C. Y. Tzen, R. E. Scott, and L. R. Johnson. 1993. Decreased expression of protooncogenes c-fos, c-myc, and c-jun following polyamine depletion in IEC-6 cells. *Am J Physiol* 265: G331–G338.

Wang, J. Y., M. J. Viar, and L..R. Johnson. 1994. Regulation of transglutaminase activity by polyamines in the gastrointestinal mucosa of rats. *Proc Soc Exp Biol Med* 205: 20–28.

Wang, Y., L. Xiao, A. Thiagalingam, B. D. Nelkin, and R. A. Casero, Jr. 1998. The identification of a cis-element and a trans-acting factor involved in the response to polyamines and polyamine analogues in the regulation of the human spermidine/spermine N^1-acetyltransferase gene transcription. *J Biol Chem* 273: 34623–34630.

Wang, Y., W. Devereux, T. M. Stewart, and R. A. Casero, Jr. 1999. Cloning and characterization of human polyamine–modulated factor-1, a transcriptional cofactor that regulates the transcription of the spermidine/spermine *N*(1)-acetyltransferase gene. *J Biol Chem* 274: 22095–22101.

Wang, Y., W. Devereux, P. M. Woster, and R. A. Casero, Jr. 2001a. Cloning and characterization of the mouse polyamine-modulated factor-1 (mPMF-1) gene: An alternatively spliced homologue of the human transcription factor. *Biochem J* 359: 387–392.

Wang, Y., W. Devereux, P. M. Woster, T. M. Stewart, A. Hacker, and R. A. Casero, Jr. 2001b. Cloning and characterization of a human polyamine oxidase that is inducible by polyamine analogue exposure. *Cancer Res* 61: 5370–5373.

Wang, Y., W. Devereux, T. M. Stewart, and R. A. Casero, Jr. 2002. Polyamine-modulated factor 1 binds to the human homologue of the 7a subunit of the *Arabidopsis* COP9 signalosome: Implications in gene expression. *Biochem J* 366: 79–86.

Wang, X., G. Shi, Q. Xu, and J. Hu. 2007. Exogenus polyamines enhance copper tolerance of *Nymphoides peltatum*. *J Plant Physiol* 164: 1062–1070.

Wang, J., P. P. Sun, C. L. Chen, Y. Wang, X. Z. Fu, and J. H. Liu. 2011. An arginine decarboxylase gene PtADC from *Poncirus trifoliata* confers abiotic stress tolerance and promotes primary root growth in *Arabidopsis*. *J Exp Bot* 62: 2899–2914.

Wang, H., J. Gao, W. Li et al. 2012. Pph3 dephosphorylation of Rad53 is required for cell recovery from MMS-induced DNA damage in *Candida albicans*. *PLoS One* 7: e37246.

Wang, X., N. A. Stearns, X. Li, and D. S. Pisetsky. 2014a. The effect of polyamines on the binding of anti-DNA antibodies from patients with SLE and normal human subjects. *Clin Immunol* 153: 94–103.

Wang, X., W. Ying, K. A. Dunlap et al. 2014b. Arginine decarboxylase and agmatinase: An alternative pathway for de novo biosynthesis of polyamines for development of mammalian conceptuses. *Biol Reprod* 90: 1–15.

Watanabe, S., K. Kusama-Eguchi, H. Kobayashi, and K. Igarashi. 1991. Estimation of polyamine binding to macromolecules and ATP in bovine lymphocytes and rat liver. *J Biol Chem* 266: 20803–20809.

Watanabe, M., D. Watanabe, and S. Kondo. 2012. Polyamine sensitivity of gap junctions is required for skin pattern formation in zebrafish. *Sci Rep* 2: 473: 1–5.

Watson, M. B., K. K. Emory, R. M. Piatak, and R. L. Malmberg. 1998. Arginine decarboxylase (polyamine synthesis) mutants of *Arabidopsis thaliana* exhibit altered root growth. *Plant J* 13: 231–239.

Weiger, T. M., T. Langer, and A. Herman. 1998. External action of di- and polyamines on maxi calcium-activated potassium channels: An electrophysiological and molecular modeling study. *Biophys J* 74: 722–730.

Weiner, C. P., R. G. Knowles, L. D. Stegink, J. Dawson, and S. Moncada. 1996. Myometrial arginase activity increases with advancing pregnancy in the guinea pig. *Am J Obstet Gynecol* 174: 779–782.

Weir, B. A. and M. P. Yaffe. 2004. Mmd1p, a novel, conserved protein essential for normal mitochondrial morphology and distribution in the fission yeast *Schizosaccharomyces pombe*. *Mol Biol Cell* 15: 1656–1665.

Wen, L., J. K. Huang, and P. J. Blackshear. 1989. Rat ornithine decarboxylase gene. Nucleotide sequence, potential regulatory elements, and comparison to the mouse gene. *J Biol Chem* 264: 9016–9021.

West, H. M. and D. R. Walters. 1989. Effects of polyamine biosynthesis inhibitors on growth of *Pyrenophora teres*, *Gaeumannomyces graminis*, *Fusarium culmorum* and *Septoria nodorum in vitro*. *Mycol Res* 92: 453–457.

White, M. W., C. Degnin, J. Hill, and D. R. Morris. 1990. Specific regulation by endogenous polyamines of translational initiation of *S*-adenosylmethionine decarboxylase mRNA in Swiss 3T3 fibroblasts. *Biochem J* 268: 657–660.

Whitney, P. and D. R. Morris. 1978. Polyamine auxotrophs of *Saccharomyces cerevisiae*. *J Bacteriol* 134: 214–220.

Willert, E. K. and M. A. Phillips. 2008. Regulated expression of an essential allosteric activator of polyamine biosynthesis in African trypanosomes. *PLoS Pathog* 4: e1000183.

Willert, E. K., R. Fitzpatrick, and M. A. Phillips. 2007. Allosteric regulation of an essential trypanosome polyamine biosynthetic enzyme by a catalytically dead homolog. *Proc Natl Acad Sci U S A* 104: 8275–8280.

Williams, K. 1997a. Modulation and block of ion channels: A new biology of polyamines. *Cell Signal* 9: 1–13.

Williams, K. 1997b. Interactions of polyamines with ion channels. *Biochem J* 325: 289–297.

Williams, K. 2009. Extracellular modulation of NMDA receptors. In *Biology of the NMDA Receptor*, ed. A. M. Van Dongen. Boca Raton, FL: CRC Press.

Williams, L. J., G. R. Barnett, J. L. Ristow, J. Pitkin, M. Perriere, and R. H. Davis. 1992. Ornithine decarboxylase gene of *Neurospora crassa*: Isolation, sequence, and polyamine-mediated regulation of its mRNA. *Mol Cell Biol* 12: 347–359.

Williams, J. R., R. A. Casero, and L. E. Dillehay. 1994. The effect of polyamine depletion on the cytotoxic response to PUVA, gamma rays and UVC in V79 cells in vitro. *Biochem Biophys Res Commun* 201: 1–7.

Williams-Ashman, H. G. and A. Scheone. 1972. Methylglyoxalbis-(guanylhydrazone) as a potent inhibitor of mammalian and yeast *S*-adenosylmethionine decarboxylases. *Biochem Biophys Res Commun* 46: 288–295.

Williams-Ashman, H. G. and Z. N. Canellakis. 1979. Polyamines as physiological substrates for transglutaminases. *Perspect Biol Med* 22: 421–453.

Winter, J. N., P. S. Ritch, S. T. Rosen, M. M. Oken, J. M. Wolter, P. H. Wiernik, and M. J. O'Connell. 1990. Phase II trial of methylglyoxal-bis(guanylhydrazone) (MGBG) in patients with refractory multiple myeloma: An eastern cooperative oncology group (ECOG) study. *Cancer Invest* 8: 143–146.

Wolff, E. C., M. H. Park, and J. E. Folk. 1990. Cleavage of spermidine as the first step in deoxyhypusine synthesis. The role of NAD. *J Biol Chem* 265: 4793–4799.

Wolff, E. C., S. B. Lee, and M. H. Park. 2011. Assay of deoxyhypusine synthase activity. In *Polyamines. Methods and Protocols*, eds. A. E. Pegg and R. A. J. Casero, 195–205. New York: Humana Press.

Woriedh, M., I. Hauber, A. L. Martinez-Rocha et al. 2011. Preventing *Fusarium* head blight of wheat and cob rot of maize by inhibition of fungal deoxyhypusine synthase. *Mol Plant Microbe Interact* 24: 619–627.

Wu, C. 1997. Chromatin remodeling and the control of gene expression. *J Biol Chem* 272: 28171–28174.

Wu, Q.-S. and Y.-N. Zou. 2009. The effect of dual application of arbuscular mycorrhizal fungi and polyamines upon growth and nutrient uptake on trifoliate orange (*Poncirus trifoliata*) seedlings. *Not Bot Hort Agrobot Cluj* 37: 95–98.

Wu, T., V. Yankovskaya, and W. S. McIntire. 2003. Cloning, sequencing, and heterologous expression of the murine peroxisomal flavoprotein, N^1-acetylated polyamine oxidase. *J Biol Chem* 278: 20514–20525.

Wu, D., S. Ji, Y. Wu, Y. Ju, and Y. Zhao. 2007. Desing, synthesis, and antitumor activity of bile acid–polyamine–nucleoside conjugates. *Bioorg Med Chem Lett* 17: 2983–2986.

Wu, H., J. Min, Y. Ikeguchi et al. 2007. Structure and mechanism of spermidine synthase. *Biochemistry* 46: 8331–8339.

Wu, H., J. Min, H. Zeng et al. 2008. Crystal structure of human spermine synthase: Implications of substrate binding and catalytic mechanism. *J Biol Chem* 283: 16135–16146.

Wu, Q.-S., Y.-H. Peng, Y.-N. Zou, and C.-Y. Liu. 2010. Exogenous polyamines affect mycorrhizal development of *Glomus mosseae*-colonized citrus (Citrus tangerine) seedlings. *Sci Asia* 36: 254–258.

Xiao, L. and J. Y. Wang. 2011. Posttranscriptional regulation of gene expression in epithelial cells by polyamines. *Methods Mol Biol* 720: 67–79.

Xiao, L., J. N. Rao, T. Zou et al. 2007. Polyamines regulate the stability of activating transcription factor-2 mRNA through RNA-binding protein HuR in intestinal epithelial cells. *Mol Biol Cell* 18: 4579–4590.

Xiao, Y., D. E. McCloskey, and M. A. Phillips. 2009. RNA Interference-mediated silencing of orthinine decarboxylase and spermidine synthase genes in *Trypanosoma brucei* provides insight into regulation of polyamine biosynthesis. *Eukaryot Cell* 8: 747–755.

Xiong, H., B. A. Stanley, B. L. Tekwani, and A. E. Pegg. 1997. Processing of mammalian and plant *S*-adenosylmethionine decarboxylase proenzymes. *J Biol Chem* 272: 28342–28348.

Yamaguchi, K., Y. Takahashi, T. Berberich et al. 2006. The polyamine spermine protects against high salt stress in *Arabidopsis thaliana*. *FEBS Lett* 580: 6783–6788.

Yamamoto, S., T. Imamura, K. Kusaba, and S. Shinoda. 1991. Purification and some properties of inducible lysine decarboxylase from *Vibrio parahaemolyticus*. *Chem Pharm Bull* 39: 3067–3070.

Yamamoto, D., K. Shima, K. Matsuo et al. 2010. Ornithine decarboxylase antizyme induces hypomethylation of genome DNA and histone H3 lysine 9 dimethylation (H3K9me2) in human oral cancer cell line. *PLoS One* 5: e12554.

Yanai, I., Y. I. Wolf, and E. V. Koonin. 2002. Evolution of gene fusions: Horizontal transfer versus independent events. *Genome Biol* 3: research0024.

Yang, H., G. Shi, H. Wang, and Q. Xu. 2010. Involvement of polyamines in adaptation of *Potamogeton crispus* L. to cadmium stress. *Aquat Toxicol* 100: 282–288.

Yao, J., D. Zadworny, U. Kuhnlein, and J. F. Hayes. 1995. Molecular cloning of a bovine ornithine decarboxylase cDNA and its use in the detection of restriction fragment length polymorphisms in Holsteins. *Genome* 38: 325–331.

Yao, J., D. Zadworny, S. E. Aggrey, U. Kuhnlein, and J. F. Hayes. 1998. Bovine ornithine decarboxylase gene: Cloning, structure and polymorphisms. *DNA Seq* 8: 203–213.

Yao, M., A. Ohsawa, S. Kikukawa, I. Tanaka, and M. Kimura. 2003. Crystal structure of hyperthermophilic archeal initiation factor 5A: A homologue of eukaryotic initiation factor 5A(eIF-5A). *J Biochem* 133: 75–81.

Yerlikaya, A. and B. A. Stanley. 2004. *S*-adenosylmethionine decarboxylase degradation by the 26 S proteasome is accelerated by substrate-mediated transamination. *J Biol Chem* 279: 12469–12478.

Yoo, T. H., C. J. Park, B. K. Ham, K. J. Kim, and K. H. Paek. 2004. Ornithine decarboxylase gene (CaODC1) is specifically induced during TMV-mediated but salicylate-independent resistant response in hot pepper. *Plant Cell Physiol* 45: 1537–1542.

Yoshida, M., D. Meksuriyen, K. Kashiwagi, G. Kawai, and K. Igarashi. 1999. Polyamine stimulation of the synthesis of oligopeptide-binding protein (OppA). Involvement of a structural change of the Shine-Dalgarno sequence and the initiation codon AUG in oppa mRNA. *J Biol Chem* 274: 22723–22728.

Zaletok, S., N. Alexandrova, N. Berdynskykh et al. 2004. Role of polyamines in the function of nuclear transcription factor NF-kB in breast cancer cells. *Exp Oncol* 26: 221–225.

Zamora, D., J. T. Melendez, and A. Galaleldeen. 2014. In pursuit of mammalian polyamine oxidase crystal structure. *FASEB J* 28: Suppl. 774.

Zarb, J. and D. R. Walters. 1994. The effects of polyamine biosynthesis inhibitors on growth, enzyme activities and polyamine concentrations in the mycorrhizal fungus *Laccaria proxima*. *New Phytol* 126: 99–104.

Zarb, J. and D. R. Walters. 1995. Polyamine biosynthesis in the ectomycorrhizal fungus *Paxillus-involutus* exposed to zinc. *Lett Appl Microbiol* 21: 93–95.

Zarytova, V. F. and A. S. Levina. 2011. Polyamine-containing DNA fragments. In *Advances in Chemistry Research*, Vol. 4, ed. J. C. Taylor, 1–54. New York: Nova Science Publishers.

Zhang, C. and Z. Huang. 2013. Effects of endogenous abscisic acid, jasmonic acid, polyamines, and polyamine oxidase activity in tomato seedlings under drought stress. *Sci Hort* 159: 172–177.

Zhang, W., B. Jiang, W. Li, H. Song, Y. Yu, and J. Chen. 2009. Polyamines enhance chilling tolerance of cucumber (*Cucumis sativus* L.) through modulating antioxidative system. *Sci Hort* 122: 200–208.

Zhao, H. and H. Yang. 2008. Exogenous polyamines alleviate the lipid peroxidation induced by cadmium chloride stress in *Malus hupehensis* Rehd. *Sci Hort* 116: 442–447.

Zhao, Y., H. Du, Z. Wang, and B. Huang. 2010. Identification of proteins associated with water-deficit tolerance in C4 perennial grass species, *Cynodon dactylon* X *Cynodon transvaalensis* and *Cynodon dactylon*. *Physiol Plant* 141: 40–55.

Zheliaskova, A., S. Naydenova, and A. G. Petrov. 2000. Interaction of phospholipid bilayers with polyamines of different length. *Eur Biophys J* 29: 153–157.

Zherebilo, O. E., N. Kucheryava, R. I. Gvozdyak, D. Ziegler, M. Scheibner, and G. Auling. 2001. Diversity of polyamine patterns in soft rot pathogens and other plant-associated members of the *Enterobacteriaceae*. *Syst Appl Microbiol* 24: 54–62.

Zhu, C. M. and H. R. Henney, Jr. 1990. DNA methylation pattern during the encystment of *Physarum flavicomum*. *Biochem Cell Biol* 68: 944–948.

Zhu, Q., Y. Huang, L. J. Marton, P. M. Woster, N. E. Davidson, and R. A. Casero. 2012. Polyamine analogues modulate gene expression by inhibiting lysine-specific demethylase 1 (LSD1) and altering chromatin structure in human breast cancer cells. *Amino Acids* 42: 887–898.

Zonia, L. E., N. E. Stebbins, and J. C. Polacco. 1995. Essential role of urease in germination of nitrogen-limited *Arabidopsis thaliana* seeds. *Plant Physiol* 107: 1097–1103.

Index

Note: Page numbers ending in "f" refer to figures. Page numbers ending in "t" refer to tables.

R

S

T - #0437 - 071024 - C4 - 234/156/9 - PB - 9780367377106 - Gloss Lamination